U21 Saves the Dardanelles

Lt-Cmdr. Otto Hersing

The Great War at Sea Series

Volumes in the Great War at Sea Series

V188 My Torpedo Boat War Missions. Friedrich von Callisen
The Submarine War 1914-1918. Andreas Michelsen
The Last Cruise of the Majestic. George Goodrich
Two Lone Ships. Georg Kopp
U-boats Westward!. Ernst Hashagen
Six Years of Submarine Cruising. Johannes Spiess
German Submarines and their Warfare 1914-1918. Albert Gayer
U21 Saves the Dardanelles. Otto Hersing

Cover artwork: Willy Stöwer's representation of U21 at Constantinople

U21 Saves the Dardanelles

Lt-Cmdr. Otto Hersing

Translated by Celestino Corraliza

Published by Trident Publishing
Miami, Fl USA

Library of Congress Cataloging-in-Publication Data

Originally published Rettet die Dardanellen

U21 Saves the Dardanelles
ISBN-13: 978-0-9963157-6-0
World War, 1914-1918 -- Personal narratives, German.
World War, 1914-1918 -- Naval operations.
Military operations, Naval--Submarine.

Printed in the United States of America

Translator's Forward

Otto Hersing (1885-1960) is one of the most famous submarine officers in naval history. He is credited with the first successful attack using a submarine launched self-propelled torpedo and later sank two British heavy warships off Gallipoli in 1915 providing critical support to Germany's Turkish ally embattled there. He operated in the North Sea, British home waters and the Mediterranean Sea. His operations spanned the war's commencement to the unrestricted submarine war phase; conducting 21 patrols over a three year span with the same submarine, his beloved U21, sinking 40 warships and merchant ships.

Hersing was a wartime hero in Germany and his fame continued postwar. He retired from naval service in 1924, penning "U21 Saves the Dardanelles" in 1932. It is a brief work considering the vast extent of his experiences in the Great War. There is no patrol by patrol narrative and review. The chapter examples reflect what he considered necessary to inform the reader regarding achievements and experiences in the naval and submarine service. It is a series of vignettes and opinion pieces. Hersing is a taciturn writer compared to the lengthy recreation of some of his noteworthy patrols in the 1928 "Raiders of the Deep".[1] Although brief the current text provides insight into his thoughts.

The text is translated in its entirety. Footnotes have been added to supplement or add clarity to the text. Illustrations from the original work did not lend themselves to duplication. Substitutes have been incorporated in the work at the end of chapters where relevant.

C.Corraliza

---

[1]A dramatic anthology of German submarine commander experiences assembled by celebrated professional writer Lowell Thomas.

Otto Hersing

Table of Contents

I
I launch the First Live Submarine Torpedo Shot in World History

The first English cruiser sinks to the ocean floor
- A terrible blow shakes the submarine -
Denied the Iron Cross

September 5, 1914.

A bright autumn day, the sea a gray-green glass pane. On the coast a smoke shrouded city mass. Chaotic harbor: Edinburgh.

U 21 lurked a full day at the Firth of Forth. No steamer came its way.

Nothing.

It strained the nerves, this futile lurking. Straining even more at a submarine captain's honor. What was the submarine's military usefulness? Had one of the boats even made a shot?

It really tugged at one's sense of honor, this waiting ...

A patrol boat pushes white bow waves in front of it, a schooner is preparing ...

There-! Smoke on the horizon. Thicker, black, more promising!

I didn't take my eyes off my glasses. Prey?

But the cruiser did not come out: it was an enemy cruiser. We recognized him by his masts, his funnels.

I evaluate like a quick calculator. Should I give up my safe port position? Dare to attack? At this distance?

That would have been nonsense. And suddenly something else occurred to me: namely that it was Saturday.

Weekend.

No Englishman celebrates a fine weekend outside on the high seas. Much more pleasant and fine if it is in the port. Even in war. And I knew, knew for sure, without any doubt: the Englishman would still come. It wouldn't be long. I just had to steer towards him ...

Nevertheless, the hours crawled like lazy dogs. If he does not come? Then the commanders of the ocean-going torpedo boats and destroyers continued to smile, smiled indulgently, and told each other the truism over a wine glass: "The submarine is such a peacetime frivolity, as our Tirpitz rightly said ..."

The cruiser was gone. Not the faintest smoke sprang across the horizon.

Suddenly masts ... a funnel ... the cruiser approaches!

He really and truly comes. An electric current filled the boat. Every heart pounded. Every brain feverish. The enemy approaches.

And in front of us he stood clearly, clearly within reach: the thought that this time it had to be done. We had to!

I checked my watch. It was exactly 4 o'clock when I started the attack.

Closer to the target ... closer ...

Suddenly a fright. Half joy, half horror spreads: this was to be the first live torpedo shot launched from a submarine!

Nobody knew how the submarine would react when an explosive device as was in a torpedo warhead exploded at such close range.

Would the boat hold tight? Would seals blow out? Would the diving tanks and oil bunkers stay watertight, and above all: how will the rechargeable battery react?

But I didn't linger long with all these considerations; I wanted to get the best shot and launch my torpedo. What would come next was a second question that I didn't care about for the time being.

First I wanted to load both bow tubes and fire two torpedoes, so if one failed or didn't run, at least the other reached its destination.

Then my helmsman Godau, who was standing next to me in the tower, said: “Lieutenant[2], one torpedo is sufficient for this cruiser; it will hit, our torpedoes are in good shape and the Lieutenant is already the best torpedo shooter in the flotilla in peacetime."

The man touched my ambition: I prepared only one tube.

The attack was setting up nicely, the boat steered perfectly underwater and everything was now waiting for the moment to command the torpedo’s release. The cruiser was getting closer, soon he was in the crosshairs ... at exactly 4:45 in the afternoon I pressed the electric button in the tower.

The torpedo was ejected with compressed air, left the tube and now shot towards its target. Of course I immediately retracted the periscope so as not to draw the opponent's attention to me.

Now we sat there in our iron coffin and waited. Would we hit? And when we hit:

Would our boat hold? Would it be shattered into pieces by the detonation of such a huge mass of explosives within our close proximity...?

Would we see the golden sun shining again in our lives?

The seconds dripped hard and slow like syrup ... ten seconds ... nothing.

Twenty...

Thirty...

---

[2] USN parlance. Kapitänleutnant in German. Hersing’s memory is faulty as he was promoted to this rank in December 1914. He was a Kapitänleutnant for the remainder of the war.

Damn! Now it must..

Nothing happened, nothing at all.

Half a minute had passed. The torpedo should have reached the ship's side long ago ...

After a minute I knew it: the torpedo had run past. We sat deeply depressed ... meanwhile the last grain of hope poured like sand in an hourglass.

How could it all have happened!? How!? I thought I had calculated with the utmost precision.

I checked the clock: 1 minute and 15 seconds had passed since the shot, so the torpedo had failed. Undoubtedly an unfathomable misfortune ...

Then the world shook!

A terrible, unimaginable detonation tossed the boat like a ball.

My first thought, as soon as I came to my senses: Did the boat hold? Reports!

"Everything tight!"

Thank God ...

The battery?

Also undamaged. Everything was ok. Now I dared to look at the effects of this horrific detonation on the enemy.

I extended the periscope.

Nothing more to see from the cruiser.

According to English news reports, I hit the cruiser under the front funnel where an ammunition magazine had blown up. The foremost funnel few through the air intact somewhere into the sea. The cruiser broke apart, the foredeck detached from the ship and immediately sank. The aft section reared up by the stern, turned over and also went under. Of the entire crew (250 men), 11 men were rescued by English destroyers and patrol vessels.

It was the English cruiser "Pathfinder", which was destroyed in 3-1/2 minutes on September 5, 1914. It was the first live torpedo shot from a submarine. As well as the first shot in world history to destroy a warship.

When I arrived in Heligoland on Monday evening, there was great jubilation among the entire island personnel. Both on the Oberland and on the Unterland[3], everywhere was crowded and shouted "Hurrah!", because I had already announced my success on the way by radio message.

But my joy that I would be awarded the Iron Cross for this success was unfortunately somewhat tarnished. After I informed my flotilla chief the entire course of events of my undertaking, he said gently: "You did a great job, but unfortunately I cannot submit you for the Iron Cross."[4]

“???"

"... because according to regulations you must have been under enemy fire. You reported yourself that the cruiser did not shoot at you."

At first I was speechless for a moment and then I replied: "Excuse me Captain; I have so far understood the nature of a submarine is I attempt to approach the enemy unseen in order to fire my torpedo. Once the enemy sees me, he either turns to avoid the torpedo, or he turns towards me and runs me over, or annihilates me with his artillery. "

However, I was quickly relieved of the worry of how to earn the Iron Cross: the next morning at 6:00 a telegram arrived from headquarters that His Majesty had awarded me the Iron Cross 2nd class. However, for my crew, who had done their duty and performed their duty as the commander had during this voyage, I had to submit several reports before they too received the E.K.II. At that time, even with the sinking of an enemy cruiser, one was still very economical with awarding Iron Crosses.

---

[3] The two districts of Heligoland, the natural plateau and manmade basin respectively.

[4] Iron Cross was awarded in two classes: Iron Cross 2nd Class (Eisernes Kreuz 2. Klasse, or EKII) and Iron Cross 1st Class (Eisernes Kreuz 1. Klasse, or EKI)

The official Naval Staff work "War at Sea 1914-1918" reports on this event in "North Sea Volume 2" on p. 7[5]:

> U 21 remained at sea during the night, on September 5 at 2:00 am a steamer dimmed down to the side lights passed behind his stern, slowly moving up and down and apparently belonging to the security system. At 4.10 a.m., other dimmed vessels were in sight abeam and ahead. At 4.20 a.m. Kapitänleutnant Hersing began a submerged approach which led him to a point 5 nm south-south-east of May Island at 8.20 a.m. Surfacing here to charge the battery, he spotted a cruiser with a mast and 2 funnels in the southeast, which he tried to approach submerged, but it soon disappeared under the horizon on a northwest heading. It wasn't until 4:00 p.m. when the cruiser again traveled towards the submarine, which had dived several times due to destroyers and fish steamers, that the attack could be immediately started. Kapitänleutnant Hersing briefly reports on the other events:
>
> 4.45 p.m. Torpedo released, hit. Went to 20 m.[6]
> 5.00 p.m. Went to 10 m, nothing more to see from the cruiser. 6 destroyers and several fishing steamers come at high speed to the sinking point. Back to 20 m.
> 8.00 p.m. Course 80 degrees. Went to 10 m.
> 10.00 p.m. Battery charged, speed 9 knots. Course 80 degrees return voyage.

[5] Otto Groos, Der Krieg zur See, Der Krieg in den Nordsee vol. 2 E.S. Mittler: Berlin,1922.

[6] At 1545 hrs [time difference is due to English and German time zones] HMS Pathfinder's lookouts spotted a torpedo wake heading towards the starboard bow at a range of 2,000 yards. The officer of the watch gave orders for the starboard engine to be put astern and the port engine to be set at full ahead while the wheel was fully turned in an attempt to take avoiding action. At 1550 the torpedo detonated beneath the bridge.

The shock of the torpedo's detonation was quit violent but not dangerous to the boat at a distance of about 1500 m in 20 m of water. The enemy began chasing the submarine with several destroyers after the cruiser's sinking.

As it turned out later the ship destroyed by U 21 was the protected cruiser Pathfinder, 3000 tons in size and launched in 1904, the lead ship of the Firth of Forth's patrol flotilla. The wounded commander was among the few survivors. According to him, the torpedo detonated ammunition resulting in the ship sinking in 4 minutes.[7]

---

[7] Extract of Captain Francis Martin Leake's report follows: It was Saturday afternoon at 3.45, we had sent the TBDs off on various errands and were returning from a sweep out to sea to investigate shipping, etc. I had just left the bridge and was in my after cabin standing by the table when the screws began to stop. I started a bolt to see what it was, but before I got away from the table, she gave a veritable stagger and tremble and everything movable came tumbling down. I got up the ladder pushed the hatch cover up (it had come down), then got the boy (my valet) out, and had a look round.

Every sort of thing was in the air. Shell room forward seemed still to be going up. The torpedo got us in our forward magazine and evidently sent this up, thereby killing everyone forward. Her upper deck was flush with the water forward and it was only a question of how long she would float. Both our cutters were smashed up, the whaler was whole so all that could be done was to get this boat out and throw all floatable matter over. A badly hurt man was brought after and put on the Q.D. While this was going on she began decidedly to go down by the bows and the 1st lieutenant gave the order for jumping overboard, he judged this very well. Personally I stayed too long and found myself on the after shelter deck with the ship rapidly assuming an upright position. I decided to stand on the searchlight stand and take my chance. This soon went under and self as well, come up again ship still there, had another dive and then got shot right clear. The situation then developed, an oar came along and then a blue jacket. Then another oar and another blue jacket. Looked for ship found her still on her nose (probably on the bottom) she then fell over and disappeared, leaving a mass of wreckage all round, but I regret a very few men amongst it, for at the time they were all asleep on the mess decks and the full explosion must have caught them, for no survivors came from forward.

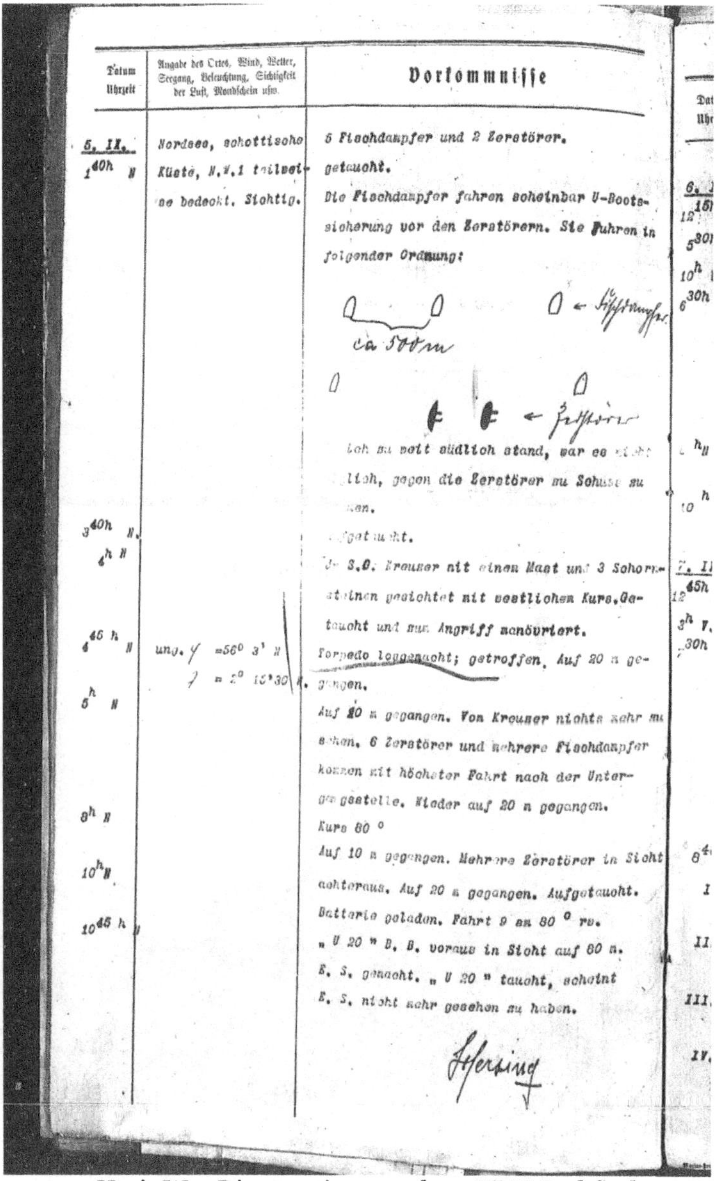

Datum Uhrzeit	Angabe des Ortes, Wind, Wetter, Seegang, Beleuchtung, Sichtigkeit der Luft, Mondschein usw.	Vorkommnisse
5. II.	Nordsee, schottische	5 Fischdampfer und 2 Zerstörer.
1^40h N	Küste, N.W.1 teilweise bedeckt. Sichtig.	getaucht. Die Fischdampfer fahren scheinbar U-Boots-sicherung vor den Zerstörern. Sie fuhren in folgender Ordnung: ← Fischdampfer; ca. 500 m; ← Zerstörer. …ich zu weit südlich stand, war es …lich, gegen die Zerstörer zu Schuss zu …en.
3^40h N		…fgetaucht.
4^h N		… S.O. Kreuzer mit einem Mast und 3 Schornsteinen gesichtet mit westlichem Kurs. Getaucht und zum Angriff manövriert.
4^45 h N	ung. φ = 56° 3' N, λ = 2° 10'30" W.	Torpedo losgemacht; getroffen. Auf 20 m gegangen.
5^h N		Auf 20 m gegangen. Vom Kreuzer nichts mehr zu sehen. 6 Zerstörer und mehrere Fischdampfer kommen mit höchster Fahrt nach der Untergangsstelle. Wieder auf 20 m gegangen.
8^h N		Kurs 80°
10^h N		Auf 10 m gegangen. Mehrere Zerstörer in Sicht achteraus. Auf 20 m gegangen. Aufgetaucht.
10^45 h N		Batterie geladen. Fahrt 9 sm 80° rw. „U 20" B.B. voraus in Sicht auf 60 m. E.S. gemacht. „U 20" taucht, scheint E.S. nicht mehr gesehen zu haben.

Hersing

U21's War Diary noting attack on HMS Pathfinder

Submarine U21

Scout Cruiser HMS Pathfinder

Otto Hersing (X) and his officers and their Iron Crosses after sinking HMS Pathfinder.

II
The Submarine in Peacetime.

A five-minute presentation for laypeople regarding the nature of submarines.

I have led my readers into medias res.[8] I took them with me on the high seas, in the middle of the atmosphere and horror of a war. I have to turn the clock back now.

What did my submarine look like? What did a submarine look like?

Its purpose? His strength? His weakness?

His brain? His armor? His weapons?

My submarine, U 21, was put into service on October 22, 1913 at what was then the Imperial Shipyard in Danzig as one of the first boats with diesel engines. With a crew of 4 officers, 3 deck officers, 10 non-commissioned officers and 19 sailors and stokers. As the commanding officer I led this boat both in peace and war and I persevered with it until the end of the war and achieved all of my successes with this boat.

I would like to provide a brief description of a submarine, above all to give a picture of its construction and equipment which enable the boat to be sailed both above and below water.

A submarine, or more precisely a submersible, is not built like an ordinary ship, which has only one outer skin, but two ships are built into one another so to speak. The inner ship is the so-called pressure hull. It is cylindrical, made of very strong steel and designed to absorb the enormous pressure at great depths of

[8] Latin phrase meaning "in the midst of things. The author commenced his narrative in the middle of the narrative without prior clarification.

water. This pressure hull now contains the entire vital facilities of a boat, such as engines, auxiliary machinery, accumulator batteries, armament and also crew living quarters (if one can even speak of "living quarters" on a submarine). A second ship is now built around this pressure hull, made of thin sheet metal, which gives the submarine the shape that is seen in illustrations when floating on the water. The space between these two ships is filled with air, and the submarine floats on this air cushion as it travels surfaced on the water. If the submarine should now dive, I only need to remove the "air cushion", i.e. I open a valve on each side at the top to let the air escape and a valve at the bottom to allow the water to flow in. The space between these two ships fills with water; the boat becomes heavier and goes deep. Underwater, the submarine is kept at a certain depth by its travel and lateral rudders[9], with which I can control any depth. Just like any airship or aircraft is held in the air by its altitude control.

During submerged travel there is nothing more to be seen of the submarine than the extended periscope, which, however, only extends below the surface to a depth of 11-12 m. This periscope is the only eye when underwater. If the submarine goes to a greater depth than 11 m there is no longer anything to be seen from the boat, even through the periscope, and the boat sails blind, so to speak.

This diving maneuver, i.e. the transition from overwater to underwater is not as easy as I just stated. One can imagine a boat about 70 m long must be exactly horizontal when the water flows into the space between the two ships. If the boat is a little lower at the front or back, which cannot be avoided when the sea is rough, it goes without saying the inflowing water first collects in the deepest end of the boat, the air rise to the other end, the boat will slowly position itself vertically and then slide to the bottom forever. To prevent this, the space between the two ships is divided into smaller sections by transverse bulkheads, the so-called diving tanks. I had 14 diving tanks on each side of U 21. If

[9] Hereafter hydroplanes as is contemporary terminology. Bow and aft respectively.

the boat was to dive, not only two valves had to be operated on each side, but 14 on each side twice. It can be seen from this that each diving maneuver was an "all-crew maneuver", because all the valves had to be operated simultaneously on the given command. Every man had to be at his post where one had to be prepared for any surprise whether the dive took place in peacetime in a home port safe harbor or outside in front of the enemy. If one or more valves did not function, the air from these diving tanks could not escape, the boat hung up in this air bubble, stood on its head and then slid into the depths.

So every dive maneuver was associated with the greatest mortal danger for the entire submarine crew. With a well-trained crew, a submarine could disappear under water in 1 minute to 1 minute 15 seconds.

If the submarine is to reappear, I do the reverse maneuver by forcing the water out of the diving tanks, i.e. I close the top valves and use compressed air, which I carry with me to 150 kg, in large steel cylinders, to press the water out of the diving tanks. The individual diving tanks fill with air, the boat obtains buoyancy and comes to the surface again.

Large diesel engines move the boat on the surface. I had two of them on board of 850 horses, each of them were coupled to a screw and gave the boat a speed of about 17 knots. However, I could not travel underwater with these diesel engines because it is well known every engine also needs good quality air to function, and this air was sorely needed for breathing underwater. So when I traveled underwater I relied on another means of propulsion. And this was electricity. Two large double dynamos were coupled between the diesel engines and the screws, and I took the power to operate these double dynamos from a large rechargeable battery, which was also installed in the boat's pressure hull. But this rechargeable battery does not last forever; the battery has to be recharged once the electricity is used up. I do this when surfaced by coupling the double dynamo as a generator to the diesel engine and stream the electrical current generated back into the battery. How long I can now travel electrically depends on how much electricity I draw at the same time from the battery. If I run at very high speed the battery will soon be exhausted (in about 1-1/2 hours). However, if I travel only in slow speed I can draw current for significantly longer, up to 38 hours. It is exactly the same as if I switch on ten lamps connected to a small battery, because then the battery will drain faster than if I only switch on one. Thus every submerged run of a submarine is connected to a small arithmetic calculation exercise: I have to think carefully about how long I want to stay underwater, how long I may or must stay underwater and how long I can proceed underwater to be able to sustain this time duration. Suppose, for example, I wanted to attack an enemy warship that was already protected by destroyers and patrol vessels, I would first have to travel submerged through this entire security to get to my target. If I succeeded in the attack, I have to have at least as much power in the battery to be able to dive under the whole security again, or even more, because in this case the enemy is aware of the actual boat and will immediately start a pursuit. The minimum I have to have in the battery is enough so I can always stay underwater until dark and then emerge under the protection of night to finally recharge my battery.

The main armament of a submarine is the torpedo weapon. A torpedo is an underwater explosive projectile that is roughly the shape of a cigar and is ejected from underwater torpedo tubes with compressed air. The torpedo then runs with its own precise machinery at a specific speed and a determined running distance towards a target in order to explode on impact and tear a large hole in the opponent's ship's bottom. Because of their length in the longitudinal direction of the boat, the discharge tubes were permanently installed in the pressure hull. I had 4 of them on U 21: 2 bow tubes and 2 stern tubes. If I want to shoot with a torpedo, I could not point the tube at the enemy as one would point a rifle and shoot at the target, but I had to take the direction of the shot with the whole boat, i.e. aim at the enemy submerged with the whole submarine. Now I cannot proceed straight to the target, because the enemy ship is in motion and the torpedo takes about 1-1/2 minutes to cover a running distance of about 2000 m. So first I have to arrive at a distance of at least 2000 m and then I have to calculate the angle that I have to maintain in front of the bow of the moving ship so the enemy ship and torpedo meet in the same place at a calculated time. The size of this lead angle naturally depends on the speed of the opponent. The greater the speed of the target, the more I have to lead him, the lower the speed the less I have to lead, and if the enemy target is immobile I can head straight for it.

As a further reinforcement, I still had two 8.8-centimeter guns on board, which were mounted on the upper deck and remained on the deck even while underwater. Only the inside of the barrel, the "breech" and the muzzle were sealed watertight so they would not be touched by the sea water.

The distribution of the entire crew during submerged travel was approximately as follows: aft with the stern torpedo tubes was the aft tube control, then the two double dynamos with their controls, then the two diesel engines and then the Central in the middle of the boat holding the watch posts. Here was the post of the diving officer and the diving control officer with his two helmsmen for the front and stern hydroplanes. Further, almost all auxiliary machines, switches, cables and pipe controls were housed here. It was the brain of the boat, so to speak. A pressure-resistant tower was built above the Central, where the commander is with navigator and helmsman. From here too the periscope was operated, with which the commander quickly took a panoramic view to make sure of his position or to start his attack.

Compressed air for the diving tanks could be operated by the commander both from the control center and from the tower. The rechargeable battery was installed in front of the control center. There were only watch posts here and then came the two bow torpedo tubes with their controls. The crew's cramped "living quarters" were located above the batteries. You had to duck your head sharply if wanted to pass through these "dormitories".

In peacetime the submarine weapon was still developing and no one knew what role it would play in a future war and what development it would undergo. As already mentioned every submarine dive maneuver was associated with great danger for the boat. Therefore, in the peacetime many precautions were taken when a submarine would dive. To mention only a few: there had to be a number of security vessels at hand at the diving site, the sea area where a submarine was diving was cordoned off, and special signals were laid down to call the submarine to surface if there was imminent danger. Then the submarines were not allowed to dive to a depth greater than 20-25 m, so that if something should happen to the submarine, divers could intervene immediately to establish a connection and save the boat crew by connecting air hoses.

Furthermore, it was easily possible to get the boat to the surface in a short time by lifting vessels and cranes at this relatively low depth of water before the crew suffocated. Of course, special precautions were taken during the first test drives of a submarine. Gradually these safety regulations were relaxed somewhat after the boats' safety and seaworthiness had been proven enduring storms during voyages. And so in May 1914, submarines were called in for a major naval maneuver in the North Sea where they were supposed to dive without security vessels for the first time. Here it became clear for the first time what the submarines were able to do in defense of the German Bight against an enemy attack or what they could do if they were to block the entrance to an enemy port. As always with such maneuvers, a contest for excellence arises between the individual boats, since every commander wants the best boat and the most efficient crew. It came about boats U 9 and U 21, Lieutenants Weddigen and Hersing, received the highest praises in the final reviews after these great maneuvers. Although submarines had proven their seaworthiness and usability on the high seas, they would have to work according to the old safety regulations in order to avoid loss of boats and human lives.

The whole situation changed in one fell swoop when war broke out in August 14. Since submarines were still not expected to undertake any major activities, our use at the beginning of the war was no different than in May 1914 during the great maneuver. We were designated as an outpost and security chain in front of the German bay and formed a large arch that began in the south on the Dutch coast at the island of Terschelling and in the north reached the Danish coast at Horns Reef. German submarines were laying here as a kind of living minefield waiting for the evil enemy to arrive. At the very beginning of the war the English declared he would get the German fleet out of its holes like rats. But the evil enemy did not come, and it was soon after mid-August we continued to expand our voyages, not only to half but across the entire North Sea up to the enemy coast.

This led to the first long-distance German submarines operations, far from their home base to the enemy coast.

1914 peacetime gathering of submarines prior to outbreak of war. U21 is in the foreground at extreme right.

III
Smothered by Heavy Shells

The fate of the submarine hangs on a thread

14 days after my first submarine success, my good comrade Otto Weddigen managed to sink the three English cruisers "Cressy", "Aboukir" and "Hogue" off the Flanders coast with his U 9 on September 22, 1914. After this second success of German submarines the Englishman withdrew all his shipping in the North Sea because the area became too dangerous for him. That is why our submarines ventured on further voyages to catch prey.

In November 1914 I drove for the first time through the Dover-Calais passage into the English Channel to the French port of Le Havre, in order to sink transport steamers enroute between England and France if there were no warships.

In the afternoon before Le Havre. The French coast somewhere ahead of me in the haze. No steamer, no cruiser, no torpedo boat. The war seemed to be taking place in the Saale[10] ...

Finally clouds of smoke. We were lurking.

A steamer pounded up, struggling hard, fully laden and laying deep in the waves.

I gave the order to fire a warning shot. It whined through the air, arching over the blue water surface and planted a giant water fountain into the sea in front of the steamer's bow.

At the time, steamers were still well behaved and the trade war was child's play. No guns hid under planks and behind bulwarks.

---

[10] i.e. far away from the current area. The Saale regon being far distant in the German rear area away from the frontlines as appeared peaceful Le Harve.

The Brit stopped immediately.

I approached him - later we couldn't dare something like that - called to him through the megaphone: “What is your cargo?”

Crowing as if from an old gramophone, From the Britisher’s command bridge echoed a resounding: "General Cargo."

Suddenly the helmsman next to me: "Fine, Lieutenant, they have General Cargo on board ..."

I ordered the captain to abandon the ship. Their boats immediately went down, the crew departed from the steamer. Then the shells whistled over, bursting into toxic gray clouds.

The steamer sank....[11]

Now one could see the French coast clearly: It is incomprehensible I was calmly allowed to sink steamers within view of the coast...

The boats of the sinking ship rowed to the land. . . now the first torpedo boats had to rush in.

They didn't rush. The afternoon passed in an undisturbed idyll.

The evening spun its first violet threads.

Suddenly there were fat, thick clouds of smoke inside it. And high-foaming bow waves that buried the entire fore-end: torpedo boats and destroyers.

Submerge ...

Through the periscope I see the boats racing towards me at a full emergency speed ... wonderful sight, this view ...

Then I also pull in my periscope, which tends to leave a telltale streak behind it. They approach me.

All shipping in front of the port has been shut down. Two days later, despite the enemy guard, I managed to sink another steamer supplying coal to France.[12] However, I soon had to

---

[11] Steamer “Malachite” of 718 tons.

[12] Steamer “Primo” of 1366 tons.

return home because submarines were not allowed to be at sea for more than eight days at the time.

Now that German submarines had also appeared in the Channel, enemy guards were also significantly strengthened here.

In January 1915 I undertook an even bigger undertaking, departing from Emden, driving west through the Strait of Dover-Calais, all the way through the English Channel, and heading north through Cape George to the Irish Sea through St. George's Channel. Here I appeared one morning just outside Liverpool Harbor, and since there were no warships either, I took on enemy steamers. The first steamer I stopped had coal loaded and was destined for the English blue water fleet in Scapa Flow. The crew was given 5 minutes to leave their ship then the steamer was sunk. It was the English "Ben Cruachan".[13] That same morning two other English steamers shared the same fate.[14] Although the entire steamers' sinking could be observed from land, it was until noon that the Englishman came out of the port with his torpedo boats and destroyers to chase me away. All navigation in the Irish Sea was suspended. Since there was no other target in sight in this area, I drove a little further north to maybe have a chance of success there.

"What can that be, Lieutenant?" I cannot truly recognize it...

Yes, what could that be, this clump of buildings in front of the Yarrow shipyard?

Suddenly it called out to me: an airship.

Then he echoed right next to me: "An airship hangar."

This was a diversion that was worth it.

---

[13] 3092 tons. The captain of the "Ben Cruachan" reported as follows: The submarine's commander expressed regret for his actions but said: 'War is war!' He expressed a hope that the crew of the *Ben Cruachan* would be picked up quickly and that the weather would not worsen too greatly. We shook hands and saluted each other.

[14] Steamers "Kilcain" 456 tons and "Linda Blanche" 369 tons.

That was at that time of the new still light hearted submarine war! I surfaced and proceeded close, or "under Land", as we say.

Started firing...

Corrected elevation and range ... there was already a black fountain of smoke and dirt. . . stood there for a long time like an ugly tree in the middle of the hanger. . .

“Two degrees to the right. . . “

Siouuuuuu. . . the projectile hissed from the barrel. This time we landed in the middle of the hanger. Then something horribly wonderful happened. A dazzling flame shot up into the sky: the hanger was blazing.

We stood erect on deck, watching the fire like a spectacle.

Suddenly two thuds.

As if I had been whipped: Those were shots.

Siouuuu. . . siouuuu. . . it arched over the water in two parabolas.

We were being shot at.

Closer, closer. . . the inevitable whimper. . .

Oh well, I think.

The next moment it is as if giant fists break a railway bridge into matches. Sky-high water fountains whoosh next to our boat at lightning speed, splinters whip around our ears, wearily tinkle on the deck. . . we are showered with water. . .

Hell, that was close.

"Dive!" I shout. . . the crew tumble below deck. . .

The next volley is already whining.

Only then did I realize - it had fallen beneath the table of consciousness in the tumult of events - the bushes near the hanger suddenly rustled, as if a hand was stroking them: there, certainly, a well masked howitzer battery had to be there . . .

We all felt it: this time it was a hair's breadth. And it was already a very thin, fine woman's hair. . .

I thought of my flotilla chief: now I had surely been under fire.[15]

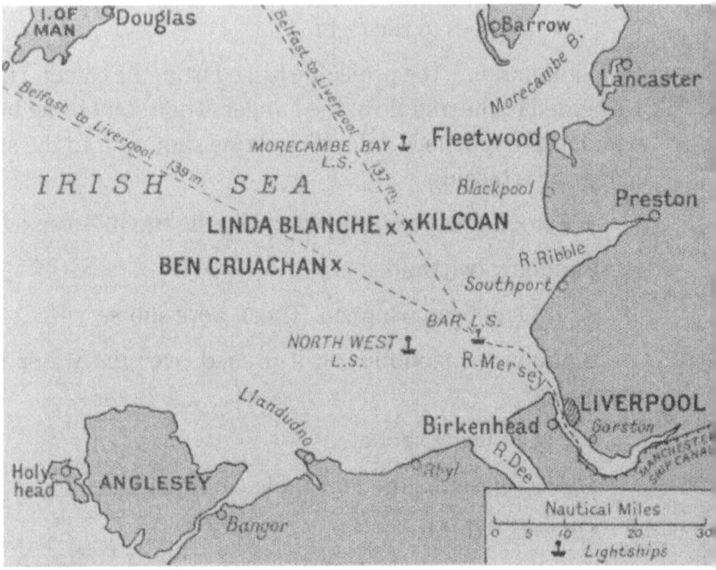

Sinking of three vessels during the cruise to the Irish Sea
Illustrated London News [London, England], 6 Feb 1915

[15] The bombardment was directed at the airship sheds on Walney Island before being driven off by artillery from Ft. Walney. Hersing commenced firing at 2.15 pm. He was sighted by the British for approximately 40minutes prior to firing and there was uncertainty regarding identification; British submarines being constructed, tested and repaired in nearby Barrow made submarine sightings a common occurrence. The engagement range was approximately 7000 yards. The fort used two 6" artillery pieces firing approximately 11 shells. Hersing fired 2 shells. British reports indicate German shells missed their target.

Steamer Linda Blanche

Steamer Ben Cruachan

IV
Surfaced through the Strait of Gibraltar surfaced in broad daylight

What does a destroyer care about a submarine when the relief arrives.

Now that I had returned to Emden from the trip to the Irish Sea, I became aware of the good material we had and what my boat could do. I had already gotten to know enough of the enemy coasts around England. I was now calculating how far I could actually get within the radius of action of my boat. My intention was to go a little further, to the Mediterranean - to Constantinople!

If everything went well, my oil supply should just barely reach Austria in the Adriatic. In March 1915, the situation down in front of the Dardanelles among our Turkish allies was as follows:

The Allied fleet, the British and French, had attempted to force free passage through the straits, the Dardanelles and the Bosphorus. They wanted to bring by this short route their food supply (especially grain) from the granary of southern Russia through the Black Sea via Constantinople which was the shortest route to France and England. Since Turkey was on our side, all of the grain in southern Russia had been left unused, and England and France had to rely much more on Canada, Australia, and Argentina to obtain their grain. This attempt to force the straits failed due to the bravery of the Turkish coastal forts.

But the enemy did not give up his operation, but tried to force the straits by land by taking the Turkish coastal forts without using warships, because the first advance had cost the

Allies four battleships.[16] The enemy had now landed on the European side at the Gallipoli peninsula and a veritable static warfare with constant trench fighting had developed at Sed ul Bar.

In these battles, our allies, the Turks, were at a disadvantage insofar as the enemy was equipped with the most modern trench warfare equipment. Unfortunately we were unable to offer any support to our allies, as the route to Constantinople via Serbia was not yet clear at the time.

While the Turks had very little artillery, the enemy was constantly lying off the coast with his large warships and taking the Turkish trenches from the flank in a murderous fire with his heavy ship guns. The presence of a submarine down there could help our allies a great deal. That's why I submitted my proposal - to attempt a voyage to the Mediterranean - to the Naval Staff in Berlin via the High Seas Fleet.

At first the noble men in Berlin didn't want to know anything about the trip at their green table.[17] They thought it was impossible.

Finally, through intercession of the fleet, I was personally ordered to Berlin in order to debate and overcome the Naval Staff's last concerns.

After several sessions I was given the command to try the risky trip to the Mediterranean. Nobody in Berlin thought that I would arrive safely in Constantinople. When I said goodbye, none of the gentlemen said "Goodbye!" to me. But instead, they packed me a large packet of documents that I had to take with me.

---

[16] HMS Goliath on May 13, HMS Ocean and Irresistible and French Bouvet on May 18, 1915.

[17] An expression meaning to debate without hope of conclusion. Drawn from the green cloth covered table believed typical in international and political negotiations in Europe of the time and preceding century.

When I asked what was written there, they told me that these were international agreements that I would have to deal with if I were forced to call at a neutral port as the warship of a belligerent.

I thanked them for it and was happy to have it in writing. I took the package under my arm and boarded the train to Wilhelmshaven. Once there, I stowed this thing in my secret locker on the submarine and did not bother myself to continue reading it.

Better instead, I took a look at the large map of Europe to find out which countries would still be neutral on my trip to Constantinople. There was England first, then France, then Spain, which was neutral. Then Portugal, then the entrance to the Mediterranean through the Strait of Gibraltar, where on one side the British sat on the rock of Gibraltar, on the other in Africa the French in Morocco. Then it went on: again France, both in Europe and in Africa, in Algiers and Tunis, then the Italians and finally a country that was neutral at the time, which was Greece.

So I didn't have much prospect of entering a neutral port. I now equipped my boat for the long journey without telling anyone where the journey was destined. No one from my crew knew either: They did not ask about it as well. At that time there was still no Soldier's Council which interfered in everything.[18] My good people knew, however, that everything would be fine wherever their commander should lead them. They also knew their commander was deadly serious, bore the responsibility and vouched for it with his own life. If this venture had become common knowledge, it would have been easy for the enemy to find and destroy a single solo submarine traveling all the way around Europe.

On April 25, 1915, I anchored in Wilhelmshaven prior to my long voyage around Europe. The nearest route would naturally

---

[18] Self-named organization of rebellious sailors, soldiers and civilians formed during the Kiel Mutiny in November 1918 who took defacto control of warships, overthrowing the conventional chain of command Hersing is accustomed to.

have been through the Channel and off Calais to the Atlantic Ocean. The other longer one is the north around England. However, I decided to use the longer one, for the following reason: I was acquainted with the trip through Dover-Calais and the Channel. Because of the presence of our submarines in the Channel, the enemy had set up such an intense guard activity that it was quite impossible for an enemy submarine to be able to proceed surfaced during daytime. I would have been constantly forced underwater during the day while driving through the Channel, then had to try to surface unseen at night to recharge my battery under the cover of darkness. I would have used a lot of fuel to charge the battery without making much progress, while on a trip north around England, off the main steamer routes I could count on getting to my destination with the most economical cruising speed. After our success on September 5 and 22, 1914, the British had already stopped shipping in the North Sea. And so I drove through the North Sea with the right April weather: heavy seas, hailstorms, rain squalls, snow flurries and fog.

When I had now arrived at the northern tip of England, in order to break through to the Atlantic Ocean, I had to reckon with colliding with the enemy for the first time on my journey, for the Englishman had already begun his hunger blockade against Germany at that time, and since it seemed too dangerous for him in the North Sea, he blocked the passage between the northern tip of England, the Shetland Islands, the Faroe Islands and Iceland by an outpost line in order to prevent a steamer from America reaching Germany.

I had to expect an encounter.

It was a real bad weather: snow gusts and fog and snow again, so that you could not see your hand in front of your eyes. The tower slowly turned into an iceberg.

Desolate this symphony of white and gray, this eternal marriage of sky and water, these wet scraps of fog that planted themselves before me in thousands of grotesque shapes.

The journey seemed endless: the outpost lines[19] did not come into view.

The weather became more and more hazy, the sea ever wilder.

Again we were stuck in a fog bank, then the flakes stormed towards the tower in whole regiments ...

How did it look to rearwards? I slowly turned ... startled. Hello: the whole chain of outposts was behind me.

Initially I thought we had slipped through in the fog and weather. Right behind me was a steamer, on the left behind me another ...

I calculated lightning fast: dive?

Nonsense. The next fog bank was already in front of me, it will swallow me up in a few minutes ...

But I vividly imagined the long faces of the British when they suddenly saw a submarine behind them in these waters ...

The steamer astern had already noticed me. He hoisted a long signal.

I continued onwards.

To shake me out of my lethargy, he drew on his steam whistle at his funnel so that it screamed eerily over the sea.

Do not ignore this warning! I remained deaf, but cast longing glances at the fog bank in front of me, which seemed to recede more and more ...

In any case, I kept a close eye on the ship, and rightly so, a sailor stomped on the forecastle of the outpost vessel, wrapped in oilskins and Southwester. What did he want?

Up forward there was an unusual mountain of gear. It was dismantled and a gun appeared. The man quickly pulled off the cover.

---

[19] Roving patrols by English vessels along a well defined area rather than a rigid line.

"Now it may be uncomfortable," I said to my watch personnel, and was glad the fog bank was finally getting closer. By the time the quick Englishman sends over his shells, we will have disappeared ...

Right, the foremost fog veils painted gray patterns around our tower.

And we were swallowed up. Afterwards I was interested in what the Brit had to say to me. I pored through the signal book and now I knew what the Englishman had asked me: Stop immediately!

However, I did not fulfill my opponent's wish because I wanted to go even further. And so I drove south into the Atlantic, west of England and Ireland, through the Bay of Biscay in very bad weather to the Spanish coast.

At this location the Naval Staff in Berlin had ordered a small steamer with oil for me through the German Embassy in Madrid, which I was to meet in the neutral waters in Corcubion Bay. The nighttime recognition and meeting with the steamer worked very well, unnoticed by the Spaniards, and I took in a lot of oil on the first night and arranged another meeting point with the steamer for the next night in a quiet bay on the Spanish coast. Before daybreak I steamed away from view of the Spanish coast and now tried the oil on this day. Unfortunately, it turned out that the freshly taken over oil was useless for my engines, because it did not ignite in my diesel engines due to a too high flash point.[20] The next night I said goodbye to the steamer without taking on more oil, and decided to continue onward. With the most economical fuel oil consumption usage[21], I should be able to reach our ally, our Austrian brother in the Adriatic Sea, where I could expect to obtain good fuel for my engines.

I drove further south and after a few days I reached the Mouse hole, the Strait of Gibraltar, the entrance to the Mediterranean. At first I was hoping to be able to cross the straits

[20] The supply ship carried crude oil instead of the required Diesel oil.

[21] About 9 knots.

surfaced at night; however, since I had to avoid several steamers in order not to be seen, I did not arrive at the narrowest point until it was dawn. On the left was the high rock of Gibraltar which dominated the entire passage, on the right I had the African coast. I didn't want to turn around and wait until it got dark because of my low fuel supply, so I tried to break through now when surfaced.

The narrowest point between the rock of Gibraltar and Ceuta Cape the Englishman had closed off by a security line of large destroyers. There were 7 or 8 torpedo boats guarding the strait by steaming up and down. In order to be seen as little as possible, I drove in all the way southwards, just off the African coast.

There I had the high dirty yellow sand dunes of Morocco as a background, otherwise the submarine easily stood out against the clear horizon.

The enemy outpost line came closer, closer. I had to be discovered at any moment.

Now, the southernmost destroyer was closest to me. I clearly saw his silhouette, I could tell from the high bow waves that he was driving north at high speed.

There was no moment to be lost: "All men at diving stations! Clear for quick diving."

My glasses were stuck to my forehead. He must have seen us long ago ... and if he didn't, he would see us as soon as he turned again.

But strangely, But strange, the English continued to drive further north to his harbor.

Suddenly there are heavy clouds of smoke coming from the direction of the port of Gibraltar. Undoubtedly, it came from vessels there the destroyer was driving towards.

I tried to and could not understand the cause.

I checked my watch. It was exactly 8 a.m. Then I had to laugh out loud: the riddle was solved!

There is such a beautiful old sailor saying that says: "The good sailor sits and dozes, he is relieved at 8 o'clock." And so it was.[22] The British went dozing further north, towards his replacement; I can still see the two vessels exchanging the watch and the new one driving down to the coast of Africa and taking good care of it by steaming back and forth so no evil enemy gets into the Mediterranean. During that interval I had long since broken through, and so I had managed to get into the Mediterranean surfaced on a clear day.

[22] Hersing is using a measure of creative license to gratify his audience. His voyage was known to British naval forces via intelligence work and Gibraltar patrols had been alerted.

V.
How I sank H.M.S. Triumph

Hounded by French destroyers - a life and death shot

Ever further through the Mediterranean. Carefully outside the main steamer lines to Sicily.

Then along the coast of this island, which they had put a brightly colored fairy-tale garden in the blazing sun: blooming cactus, blooming palm trees ... it was a godsend ...

Then we got some rain, the coast disappeared, everything drowned in the drizzle. But it seemed to be clearing up soon. It was getting lighter forward, you could see ...

Four large modern French destroyers emerged from a rain squall ...

Less than 2000 m in front of us. Directly towards us at very high speed with high-foaming bow waves.

"Quick dive!"

We got into the depths safely, but we could clearly hear the screws clanking above us.

It faded away, again came ...

When I finally dared to poke my periscope out, I saw one of the destroyers approaching me directly.

Down again.

There was no longer any thought of surfaced travel during the day. I initially tried to get further east at night off the coast of Sicily, but it was hopeless because of the pursuit by the French destroyers.

That's why I made a detour south around the island of Malta to push ahead. I managed to make good progress here, but when

I wanted to move further east and north to swing into the Adriatic Sea, the enemy came into view again.

Chased by the French, after difficult hours, I finally came to the Adriatic Sea and to our ally, and ran into the Bay of Cattaro, escorted in by Austrian torpedo boats.

The journey up to this point was a success; the machinery had persevered without failure, even if I only had just under half a ton of oil when I entered harbor.[23] The reception was excellent and warm, I had new fuel alongside in less than 24 hours, and after the engines were a bit overhauled, the onward journey to Constantinople could begin.

As peaceful and scenic as the Bay of Cattaro was, one could feel here too that war was actually being waged. The beautiful bay is overlooked by the high Mount Lovćen, which was occupied by Montenegrins. From there, the Montenegrins fired a few shells daily against the weak Austrian positions. Every day between 10 a.m. and 11 a.m. they brought out a gun from behind large rocks and fired their 8 to 10 rounds into the area; at midday the gun was brought back into cover and the war was over for that day. Nevertheless, I couldn't stand it in the harbor anymore: after a few days when the boat was overhauled and provided with provisions, I set off again to reach my final destination, the Dardanelles.

Through the southern Adriatic Sea to around the southern tip of Greece between Greece and the island of Crete, the journey continued into the Aegean Sea. There was land everywhere, partly mainland, partly islands, ancient Greek temples dazzling white in their marble. Beautiful blue sky, bright weather, the smooth sea as blue as the sky, a real relaxation trip! But I was careful! I could not know if there was an observation post somewhere on an island that immediately reported my arrival in front of the Dardanelles and incite enemy forces on me. Besides, a mirror-like sea is not attack weather for a submarine, because if the boat shows its periscope on the surface while submerged, the

[23] U21 began the journey with 56.5 tons onboard.

foam wake is immediately visible and the submarine is betrayed. Exactly as if you were walking lengthwise in a bathtub using a walking stick. The closest route would have been, of course, if I had driven directly from the south to the Dardanelles, but again I made a small detour, I drove to the West then North under the Greek coast, passed off Mount Athos and the Bulgarian coast further to the East. By coming from the north, I wanted to get between the Gallipoli peninsula and the islands of Samothrace and Imbros, before entering the Dardanelles. I came to this decision based on the following consideration: Many of great rivers flow into the Black Sea: the Danube, Don, Dniester, Prut, etc., and all this water must also flow out of the Black Sea and now flows through the Bosporus, the Sea of Marmara and the Dardanelles to the Mediterranean Sea, at a speed of about 4 knots per hour in the Straits. It goes without saying this stream continues in a southwesterly direction from the mouth of the Dardanelles into the Aegean Sea.

If I had come directly from the south, I would have had to fight against the current underwater and my battery would have been largely empty before I even got to the enemy. While if I came so quietly from the North and after a possible success I could drift south with the current. Thus, on my detour west of Dedeagach, I came across the first enemy outpost line, which was laid out between the island of Samothrace and the mainland. The enemy seemed colossally wary to already post his outposts at this distance. After I found the enemy, I turned away immediately, not out of fear, but because I wanted to wait for the night to pass the enemy outpost line surfaced in the dark.

After dark I continued my advance toward the Dardanelles estuary where the enemy's landing site was. I came through the first outpost line unseen on the surface. A little further south, however, the enemy had laid out a second line of outposts so dense that it was impossible to get through survaced. I had to dive, went to 40 m and continued south during the night with the minimum speed needed to keep the boat underwater.[24] The next morning, with daybreak approaching, I went to 10 m where I could just show the periscope to observe the area.

It was not to be as simple as I had imagined; there was a bustle of activity on the surface. Everything the enemy had in torpedo boats, patrol vehicles, outpost steamers, etc., was criss-crossing at high speed and zigzagging between the Imbros Island and the Gallipoli peninsula in search of a submarine.

Now I realized the French, who saw me eight days ago near Sicily and pursued me, had immediately reported my presence to the Dardanelles. It wasn't long before they saw my periscope and the chase began. From 6.30 in the morning until around noon I also drove south in zigzag courses to be as unseen as possible in front of the entrance where the enemy's landing site was.

Around 12 noon I suddenly discovered an enemy warship just below the coast of Gallipoli, it was the English battleship "Triumph". I immediately retracted my periscope and headed towards this ship of the line, because I had found the enemy inshore security patrols were significantly weaker. After about 3/4 hour I came back to 10 m to start my attack.

The English battleship had fired from the flank over the Turkish trenches in the morning and was now taking a lunch break. The ship traveled back and forth off the coast with torpedo protection nets deployed in place at about 5 to 6 knots. The whole crew lay on the upper deck and basked in the sun, only the lookouts, heavily armed with extra glasses, were looking for a submarine.

I immediately started my attack, but unfortunately I was still a bit too far away to be able to shoot. When I tried to retract my periscope so as not to betray my presence, I discovered something that did not suit my attack. The English battleship had finally secured itself a large destroyer for protection, which

---

[24] A submarine usually needs to maintain forward movement to remain in neutral buoyancy otherwise it descends into the depths. Exceptions depend on the type of current encountered if any.

always steamed at high speed around the battleship, at a distance of about 1000 – 1200 m – the most favorable firing distance for a submarine-which of course the British knew exactly. As I was about to retract my periscope to proceed to firing distance from the enemy, I see this destroyer coming around the battleship's bow and drive towards me at high speed.

In a split second I had to think:

Dare to attack? That meant the full commitment of the boat! If I was destroyed, it would take six weeks before another German submarine was there again ...

Why now? The Briton would come back to this position tomorrow or the day after tomorrow...

Closer! The destroyer foams up at lightning fast speed ...

After all, the mere presence of a submarine in front of the Dardanelles was enough to make the enemy uneasy ...

Just a moment more and the opportunity to attack would be over.

Then I dared!

"Go to sixteen meters!"

The destroyer could no longer ram me at this depth ... I could not go deeper ... the time it took. And I had to shoot immediately if I didn't want to collide with the "Triumph" I was headed towards ...

Now I was at 16 m, the screws of the destroyer clanged above me ...

The screws sounded louder and louder ... what was that? The boat rocked ... now water bombs would detonate ... it was bound to destroy us, because he was just above us ...

We held firm, caught our breath, because now something terrible had to happen ...

The seconds dripped slowly ... nothing.

The clank of screws subsided ... no bomb tore the boat apart. He hadn't noticed us!

"Go to ten meters!"

I look through the periscope: the battleship is 400 m in front of me. It passed the crosshairs, had to turn hard ...

Increased forward speed. 300 m ... 200 m ...

I release the torpedo there! Through the torpedo net!

I look before I go under, the only possible path is to pass under the battleship, all the guns are aimed at me ... crashing volleys.

A terrible blow occurs ...

The submarine is thrown around like a ball ... we tumble against the bulkheads ...

As we recover from our horror, I ask the compartments to report: Thank God everything is tight ... the battery intact ...

It was a horror that was worth it. The battleship capsized and disappeared in a blue deluge in 9 minutes.[25]

Now the enemy knew there was a submarine and the pursuit began immediately. I drove south as I had intended to get away from enemy patrols, hoping to be able to surface at night to recharge my battery. But the Briton stubbornly sat on my heels and only late after midnight, far to the south, could I surface. I had been under submerged continuously for 29 hours during this attack!

When I arrived in Constantinople, I spoke to a German comrade who had been lying in the trenches on the Gallipoli

[25] An officer onboard HMS Triumph reported: "... we saw the periscope on the starboard side and the trail of torpedo right after at 12.30. Instead of bursting in the nets as we expected, it went through as if there was no net there and hit us in amidships. Triumph lay 20 degrees on the starboard side and it was impossible for us to take our lifeboats. Our list quickly deteriorated, and after a few minutes it became clear that nothing could save the ship, so the order came to abandon ship. A destroyer came to the port side, but the net defense made things difficult because it was impossible to create openings on the port side, except for the upper deck and stern deck. For myself, I jumped from the upper deck and swam away from the ship. Ten minutes after the torpedo hit, the ship capsized and remained upside down for a while, eventually gradually disappearing... "

peninsula and watched the destruction of the English battleship from there. He said something like this: They had noticed something was in the air for a few days from the increased, intensive enemy patrol activity. However, no one suspected and did not want to believe a submarine would appear before the Dardanelles. When the English battleship was torpedoed and sank at noon, no one on the towering coast thought of trench warfare. Both Turks and the English and Australians who faced them climbed out of their trenches and watched the spectacle as the battleship sank at their feet.[26]

It was on May 25, 1915, exactly one month after I left Wilhelmshaven. The battleship had been stationed in East Asia at the beginning of the war and caused a lot of trouble for our coastal forts there in the fight for our Kiautschou colonies because of its heavy medium artillery.[27] Afterwards I received enthusiastic letters from the comrades who defended Kiautschou and at that time were already in Japanese captivity, regarding my sinking of "Triumph" which had so badly affected them.

---

[26] Presumably Korvettenkapitän Wilhelm Tägert who later recounted the identical event.

[27] Pre-dreadnoughts such as "Triumph" shipped more medium artillery, 14x1 7.5" vs. heavy artillery, 2x2 10" hence the term above "heavy medium artillery".

HMS Triumph

HMS Triumph listing over

VI
Destruction of the "Majestic"

The English battleship capsizes - but my boat gets caught in a net and is almost lost

Now that this attack had worked out so well for me, I decided to try the same thing again.

I drove west again the next day, sweeping north under the Bulgarian coast, and stood in the evening on May 26, north of the island of Samothrace, exactly in the same place as two days ago. The enemy outpost line was also back on the scene. I managed to get through the first line of outposts surfaced unseen, with the second I had to dive again and stayed underwater at 40 meters for the night. Nevertheless, I adjusted my speed underwater in such a manner that I stood off Cape Helles early in the morning when it was getting light and could dive during the night undisturbed under all the guarding patrol forces in the sea area between the island of Imbros and the Gallipoli peninsula. The enemy had their landing point at Cape Helles, the southern tip of the Gallipoli peninsula; I could assume there was a warship protecting the landing site to strike there. Since there was no suitable harbor with jetties for disembarking troops on the coast of Gallipoli, the Englishman had constructed a jetty in a very simple manner by running the longest possible steamer onto the beach at high speed.[28] The steamer's bow lay high and dry on land, while its stern was still swimming in the water. Here he docked his vessel and disembarkation of troops and material now went ashore via the steamer. A very simple way to build a jetty, just something very expensive and not too durable in the long

[28] The steamer "River Clyde" used as a landing ship in the initial assault at Gallipoli peninsula.

run. This landing site had to be protected by a warship and I was not wrong in my assumption.

When I showed my periscope for the first time on May 27th in the morning to orient myself (it was around 6:30 a.m.), I saw a large number of small guard vessels and steamers, which were like the wasps in front of their nest around this landing site and drove here.

But where was the warship that protected all this lively business?

I drove back and forth: saw nothing. Strange.

The masts of a warship suddenly grew beyond the many vessels. They towered up gray and mighty, at least guarded by their protective color in the colorful confusion.

According to the silhouette, it was the "Majestic" that was in these waters and had now anchored close to land, so that in the event of a torpedo hit she would not go the same way as her brother, the battleship “Triumph”, two days ago. The commander hoped in the event of being hit, his ship would presently settle on the bottom leaving the superstructure to protrude from the water so the crew could still save themselves.

It was difficult to test whether this calculation was correct. No matter how many times I drove up and down among the vessels I found no gap to put my torpedo into the body of the Englishman.

I drove down the whole nest again, when I finally found a spot.

It was not large, when I went up it was 20 m. But I tried. Aiming precisely with my whole boat, maybe ...

Now I was in a position exactly aft of the "Majestic". I pressed the electric button ... the boat shook. The shot was out.

The torpedo drew out its fine bubble trajectory, through the many vessels ... if nothing came in between, if none of these motor barges came in ...

Nothing got in the way: a world shaking detonation ...

I had apparently hit the boiler room ... suddenly I saw the ship moving ... strange ... head-on, as the sailor says ... it seemed to be about to capsize ... subsided ... a minute passed , a second, third, fourth. Then the ship turned like a whale, overturning completely, waves roared over the other small vessels, and out of the chaos, from the nameless panic, the fore-ship of the "Majestic" suddenly rose, keel-up ... the rest of the battleship disappeared.[29]

It took four and a half minutes to destroy the 15,000-ton battleship!

This time again I escaped my pursuers by driving away to the southwest using the current.

The next day I made the same circumvention once more, by holing up under the Greek coast to the North, and on the evening of May 28th I again began my advance from the island of Samothrace to the entrance of the Dardanelles.

But far and wide no enemy warship was to be seen which would have been worth a torpedo shot. Only torpedo boats and

---

[29] On board HMS Majestic Petty Officer J,G. Cowie recounts the event after alerting an officer to the presence of a periscope, who replied "'Yes, and here comes the torpedo.' The commander shouted a warning to all the men on the port side, and I shouted down the ventilator leading to the engine-room. The torpedo was traveling at a good rate, and I saw that it was going to strike somewhere forward, so I went to the end of the bridge to see the effect.

It struck! There was a great muffled roar and the old ship quivered and shook in a terrible way. The masts and yards swayed as if they were coming down on top of us, but they held good. A huge volume of water shot up for about 200 feet in the air, on the port side, and I knew that our end had come. Things began to move very quickly then. Men were running in every direction, but very quietly and swiftly, for they realised that we must fight for our live, and that at any moment an internal explosion might ruin our last chance. The order was passed 'every man for himself' and everything that would float was being commandeered as a possible aid to rescue. Everyone seemed very cool, and there was scarcely and shouting... Tugs and trawlers were already coming to our assistance, but they couldn't come alongside as the ship had heeled over a lot to port, and it would have been dangerous for them. There was only one thing to do, and that was to swim for it."

guard vessels drove up and down off the coast at high speeds. After much back and forth I discovered several clouds of smoke in the port of Imbros Island and drove there, hoping to find a target at that post. But even there were only small vessels.[30] Unfortunately, I soon established I had been discovered and the enemy now resumed my pursuit.

It was the best, I thought, to leave the place as quickly as possible.

Suddenly a fine jerk. Then a strange feeling: as if the boat was being held back. And at the same time chafing on the fore-end. I heard it, my crew heard it and looked at me with horrified faces.

"Full Astern Emergency!"

The chafing became stronger, the ship stirred a little, then it was stuck again. We were caught in a net!

I knew these nets, the finger-thick wires that lay against the ship's hull, while the buoys that carried the nets danced on the surface of the sea. The guard ships would have to be asleep to not notice. Now it could only take seconds to rain explosives ... our life was short-lived ... a few minutes more ...

The boat did not move. I tried something different.

"Flank speed ahead!"

The screws whipped the water ... the net held ... held ... we were stuck again ...

Now bombs had to ...

Then something wonderful happened. The boat twitched ... quietly ... then more ... suddenly some movement ... and suddenly with a jolt that we had to hold on: we were moving! We were through. Was that a sigh of relief?

[30] Hersing omits mentioning he believed a Queen class pre-dreadnought was inside Kephalo Bay.

But our joy was a little premature. Progress was extremely slow. We had not freed ourselves, we had only carried along a part of the net which was apparently large and were now dragging it away with us along with the steel net and buoys. I now drove with this fencing mask draped around my ship 60 m below the surface of the sea, listening breathlessly to the sound of the screws ... waiting from second to second for the cursed wire to get caught in my screws ...

Now I thought I was in an area away from patrol vessels were. I tried my rudder with fear: it was still free from the embrace of the wire ...

"Surface!"

No sign of the enemy.

After 1-1/2 hours of hard work, we managed to free the boat from the diabolical net.[31]

Another effort I made on the 30th in front of the Dardanelles presented to me the same picture as on the 29th, there were no large warships anywhere.

The entire mighty English and French fleet in the Aegean Sea had disengaged due to a small German submarine and had hid in the very sheltered harbor of Lemnos.[32] I now drove further south near the port of Budrum on the coast of Asia Minor, where I had arranged to meet with the Turkish ally. Here I wanted information regarding the entrance to the Straits and the Turkish minefields.

---

[31] Approximately 100 meters of netting attached to the boat. Nets were used to trap submarines and a design deployed from ships was to trap torpedoes. Both HMS Triumph and Majestic had anti-torpedo nets deployed; Hersing's torpedoes were fitted with net cutters projecting from the warhead.

[32] British battleships were withdrawn from bombardment duty after the double sinking.

Having learned the situation of the Turkish minefield, I went back to the Dardanelles and could still not see a hostile ship, I entered the Dardanelles on 4 June at the greatest possible water depth, where the outgoing current was not as strong and I could duck below any mines possibly laid by the enemy. Soon after noon I called at the Turkish port city of Chanak and reported to the commander of the straits, Admiral Usedom Pasha, after I had previously given my calling card outside. The next night I drove through the Sea of Marmara and arrived at Constantinople on June 5 in the morning when the sun was shining brightly on the golden domes of the mosques and minarets. I proceeded into the Golden Horn under the Galata Bridge cheered on by the Turkish people positioned on it.

This first voyage of a submarine from Wilhelmshaven around the whole of Europe to Constantinople, where the boat had sunk two enemy battleships before it arrived, despite hostile counteraction, is not only a success of the efficiency and daring of German sailors, but also a success German technology that managed to build submarines that were able to withstand such demands.

The success of this trip is also in a political and strategic area. Due to the presence of German submarines off the Dardanelles, the enemy saw the hopelessness of his operation against the straits and it slowly diminished. Because of this, Bulgaria, which until then had been neutral, entered the war on our side, Serbia was now attacked and overrun by both sides, and our Eastern Front, which at that time was interrupted by Serbia, now reached in a single line from the Baltic Sea, from the Gulf of Riga through Russia, through Poland via the Carpathians, Serbia, Bulgaria, Constantinople, through Asia Minor, Mesopotamia and Palestine to the Suez Canal.

In recognition of this cruise I received the Pour le Mérite[33] medal and my crew the E.K.I, unless they already have it.

---

[33] Hersing was the second submariner awarded the PLM. The first being Otto Weddigen, mentioned previously in the text.

It was later claimed in press reports that U 51 had made the journey to Constantinople. This does not correspond to the facts. Submarine U 51 was still under construction at the time and was not yet ready for the frontline. On the way I painted No. 51 in white paint on the bow of my boat, it was a deliberate deception of the enemy and the press, which I intended and permitted as a military stratagem, in order to deceive them regarding the exact classification of our submarines. U21 was the only boat I commanded during the entire war.[34]

---

[34] Hersing provided an account of his operations, identifying his boat as U 51, to the neutral and friendly press. An example provided to the Chicago Daily News in Constantinople after the above events transpired demonstrating the relatively censor-free naval reporting typical in the early wartime phase:

HOW HE TOOK U-51 TO CONSTANTINOPLE

"We left Wilhelmshaven April 25th," Captain Hersing said. "When I was told that I was going to Constantinople I said: "Good! That is worthwhile." I let only the first officer and the chief engineer into the secret to explain the increased supplies taken aboard. We slipped quietly out of Wilhelmshaven, as have so many other submarines going into the war area. We kept on the surface most of the time, watching for big ships of the enemy, but failed to see any for a long time, to our great disappointment.

Dive for Safety

"Off the coast of England we were finally fired upon by a destroyer and had to dive for safety, but along the French coast we encountered no enemy. When 100 miles from Gibraltar we were fired upon by other British destroyers. We waited outside Gibraltar and passed through the strait early in the morning in plain view, but without drawing a single shot or attracting the least attention. Those Were tense but gratifying hours. "Once Within the Mediterranean the watch, already exacting, was redoubled. Here again we encountered ships of the enemy near a small island, but dived to safety under fire. When passing Malta we were fired upon by a French destroyer, but were untouched. We then proceeded south of Greece into the Aegean Sea and to the Dardanelles. "We arrived there on the night preceding May 25th, having come from Wilhelmshaven in exactly one month. In the early morning light we saw the Triumph and the Majestic lying off the coast, constantly encircled by destroyers. Through the periscope I saw a destroyer coming directly for us. We dived and the destroyer passed immediately over us with a sound like that of a motor car. "We came up immediately. I took aim through the periscope and pressed the button, automatically firing the torpedo, and the projectile slipped noiselessly into the water. We dived again. The explosion which followed was as terrific as though it had been in the fore part of the submarine itself.

Two Days Under Water

"Then we lay hidden two days and a half," continued Captain Hersing, "after which we came up again in the midst of the British ships. Just before noon,

HMS Majestic
HMS Majestic leaving Mudros for a mission off Gallipoli

---

looking through the periscope, I saw the Majestic surrounded by ten ships steaming around her in a constant circle for her protection. I could see the Majestic sailors on the deck taking their noonday nap. "Shall I disturb them?' I thought. Then, seeing a welcome space between the encircling ships, I pressed the electric button and the torpedo was going, right. It caught the Majestic a little to the rear of amidships. "We dived again. We noticed that the bombardment from the ships had ceased, for they had beer; shelling the Turkish land position. "We remained submerged for several hours and then came to the surface, to find the British ships had disappeared and all search for them was in vain. We came to Constantinople, having passed forty-two days in the submarine without rest or change."

HMS Majestic sinking

HMS Majestic overturned. The hull would be visible for weeks afterwards.

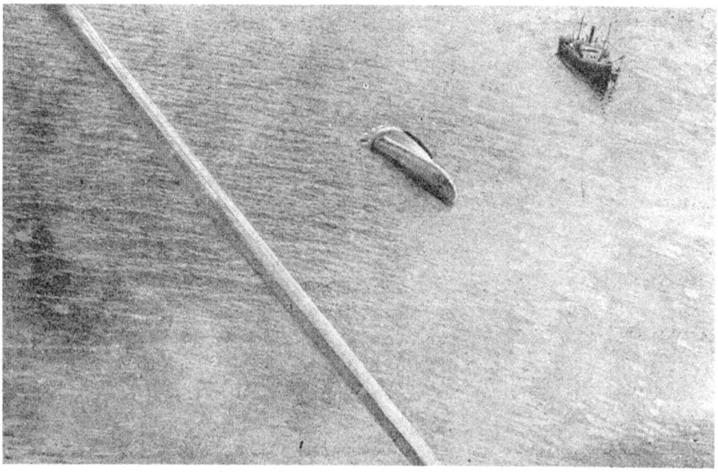

U21 at the Golden Horn in Constantinople

U21's crew at Constantinople. Hersing is front row, middle fourth from left. PLM around his neck.

Besatzung von S M. U 21 — zurückgekehrt von der großen erfolgreichen Reise
Kommandant Kapitänleutnant Hersing

VII
Ran into a Mine!

A terrible detonation - through the barrier to Constantinople with a weak battery

In Constantinople it was now above all a matter of repairing my boat as quickly as possible for further major operations. Machinery was overhauled by our own personnel; oil and food were easy to get into Constantinople. There was solely a shortage of reserve torpedoes and so I had no choice but to go on the Turkish cruiser "Midilli", the former German cruiser "Breslau", and supplement my torpedoes from there. After all, the "Breslau" had the same torpedoes on board as I did. However "Breslau" had not been able to fire an live torpedo shot since the beginning of the war at the breakthrough from Messina to Constantinople and still had all her reserve torpedoes on board. The Turks were very accommodating in overhauling the boat and machinery and helped wherever possible. I was relieved of this concern for the time being. Then I had found out while cruising in front of the Dardanelles and while entering the straits that the maneuver was not one of the easiest. Above all, it had to be expected that the enemy would do everything possible to make it more difficult for German submarines to operate in front of the Dardanelles. I therefore approached the commander of the straits with the suggestion of establishing a submarine base on the Aegean coast outside the Dardanelles. There the boats, more submarines to be added later, would be able to receive the latest news about the enemy activity, especially regarding English minelaying before entering. The base was also to be set up in such a manner so boats could take on provisions and oil in small quantities in case of an emergency.

For this purpose I now proceeded to Chanak on a Turkish torpedo boat, where accompanied by a German naval officer and several Turkish cavalrymen, I set out to find a suitable point, first on the Asiatic side. We rode west past the Turkish coastal forts to Cape Kum-Kale across from Sed ul Bar, which the British occupied, and tried to find a jetty between the Asian coast and the island of Tenedos. There was the Turkish village of Penishire, which had been completely destroyed by the English and French and abandoned by its population. As we arrived late in the evening, we spent the night as best we could in a reasonably well-preserved cellar together with feral dogs that kept bothering us in our sleep and trying to eat our provisions.

At dawn we went to the edge of the cliff, where the Turks had dug trenches to counter an enemy landing, and there I had the Turkish commander, who had followed the matter at that time with great enthusiasm, tell me more about the sinking of the "Majestic". Possibly we had shown ourselves too carelessly or the Englishman had been watching us the day before by airplane because an English monitor suddenly appeared and anchored on the very shallow water between the island of Tenedos and the mainland and began to point its only heavy gun at our observation post. The first shell hit far behind us, but the enemy had shot very well laterally[35]. After about five minutes the monitor's second shot followed, which was also good laterally and this time hit the ground just ahead of us. Now we realized the enemy was after us and we agreed it was best to leave our observation post when the next shot flashed and move out through the side trenches. After about five minutes the third shot flashed and we ran as fast as we could and the shell came howling and struck close by. After the stones had flown to the ground and the dust cloud had cleared, we wanted to go back to our observation station, but instead we found only a large crater about 10 m in diameter and 5 m deep. There was little prospect of setting up a somewhat protected submarine base at this point. Above all, the enemy could easily occupy the opposite island of

35 Meaning target bearing from the muzzle.

Tenedos. I thought it was not expedient to go further to the south to find a base because of the poor roads and connections to the interior.

So we rode back to Chanak to try our luck on the European side of the Gallipoli peninsula. However, I did not want to miss the opportunity to visit the ruins of the legendary ancient Troy. When we rode through the Scamander Plain, where the battle between Achilles and Hector was fought approximately 3000 years ago, the enemy spotted us from the opposite side of Sed ul Bar and was now starting to fire shells at us. But this did not deter us from looking further at the debris and examining it more closely under the enemy shell fire.

The next day we first visited the Turkish positions at Sed ul Bar and Kabatepe. Here, in the Turkish commander's dugout, I was obliged to take a seat in a large, comfortable deck chair, which I would have suspected anywhere else, but not here at the very front of the Turkish positions. When I asked him how he obtained this comfortable chair, he proudly showed me the name "H.M.S. Triumph "and said after the battleship sank, the deck chair drifted down the coast with many other objects.

There was no space for a submarine base right near the front, so we rode back to Liman Sanders Pasha's headquarters, to whom we reported our plans, and tried our luck the next day in the north of the peninsula Gallipoli in Saros Gulf. Here was a small sheltered bay, which was not too difficult to reach from the next inhabited place over the mountains. The mountain itself offered a wonderful view over the entire Saros Gulf to the islands of Samothrace and Imbros, so that an approach by enemy forces was very easy to see at great distances and could be communicated to the submarine below. Locating and approaching this base from the sea was very easy for those privy to it, while the enemy could only discover the small, secluded bay after a thorough search of the entire coast. A telephone was immediately placed at the observation stand from a nearby station and the observation stand itself was occupied by several German non-commissioned officers and a Turkish company. The stand itself was given the name "Hersing's Stand"[36] in my honor and served many submarines during the war.

How small the whole world is I want to briefly demonstrate with the following little experience: When I arrived at my stand in later days and was waiting for news from Constantinople, the German commander of the Turkish company asked me by signal whether he might come over to look at a submarine. I allowed it, and when he was on board, introduced himself as an engineer on the Baghdad Railway, who volunteered at the beginning of the war. But our surprise was even greater when he saw my watch officer, Lieutenant[37] Niebuhr, and greeted him as an old schoolmate. The two of them had been at school together in Hirschberg in Silesia about 14 years ago and had not seen each other since then, only to meet again in a completely deserted area of the Gallipoli peninsula during the World War.

After returning from this exploratory trip, my boat was soon ready again, so that I could think about the next operation. I ran west again through the Sea of Marmara to begin my work against the enemy warships and reinforcements in front of the Dardanelles. When I passed Chanak on the morning of July 5, I was informed a large French steamer had arrived outside the enemy landing site the night before, from which small vessels were hastily unloading war materials. This French steamer "Carthage" came straight at me and after I had passed the mine barrier, I went to 10 m and offered the steamer a cigar in the shape of a torpedo as she drove past. He couldn't handle this heavy tobacco and after five minutes he was on the sea floor, with his ammunition and his trench ordnance.[38]

---

[36] "Hersingstand" in the original.

[37] Lieutenant Junior Grade in USN parlance. Oberleutnant in German.

[38] Extract of Carthage's captain report follows: On June 30, we started unloading ammunition, the rest of the load to be unloaded at Mudros. To save water and coal, we have kept only four boilers in operation. From June 30 to July 4 the unloading was carried out with some interruptions because of the changes of anchorage made necessary to avoid the projectiles fired by the enemy batteries on the auxiliary transports at anchor.

However I was not as lucky as the first time in my further cruises in front of the Dardanelles.

I was proceeding in deep water when I suddenly heard a strange rubbing against the hull.

What was that?

The rubbing proceeded slowly along the hull. Instantly I knew with terrible certainty: the boat brushed against the anchor chain of a mine!

The English must have laid new mines in these waters unnoticed.

The seconds it took the boat to pass the rope seemed hours ...

They were not pleasant seconds, but they also went by ...

Suddenly a slight jerk. Had the rope caught on the hull?

On the contrary, it skipped off the hull, so we were saved ..

A detonation that we flung us onto the bulkhead: the mine ... the mine had detonated aft!

I called the compartments. A seal had blown out ... but ... the pressure hull seemed to have kept tight ...

That had been a shock, that had been a shock ... !!!

"The battery?"

"All in order," was the report.

But it was not accurate. Suddenly it only produced half power.

"Compressed air, Surface!"

There we could then be able to assess the damage.

On July 4 at around 1:48 pm, the cry "A submarine! Was yelled by men onboard. Finding myself on the deck, I ran aft and saw the wake of a torpedo which was only 100 meters from the side. It struck the port side of Carthage a little behind the engines at 1:50 p.m. The ship, shaken by a violent shock, instantly began to sink by the stern with rapidity, then raised its bow almost vertically, immediately fell and sank. It was 1:54 p.m.

The ship would not surface! At least not completely: it hung in the water with a great load aft.

Only by counter-flooding the front tanks could I trim the boat somewhat horizontally, however the top edge barely came out of the water. Only the tower was above water.

What to do now! The damage to the ballast tanks would not have been so dangerous if not the storage battery had not suffered so much. Trying to enter the Dardanelles against the current with only half the battery was a very risky undertaking and should have a 90 percent failure rate. I first drove to the observation post in the Gulf of Saros and reported my condition, informing I wanted to try to force the entrance to the Dardanelles in the next few days. Very few of my crew suspected the thin thread on which hung the life of the boat. Trying to get to the Dardanelles from the north, as at the end of May, was completely hopeless, because my battery did not supply as much electricity. I therefore tried the approach via another route. During the night I drove south surfaced close to land west of Samothrace and Imbros islands so as not to be seen by the enemy outpost lines. I then wanted to run surfaced under the south coast of Imbros to get as close as possible to the Dardanelles and then dive at dawn to be able to master the trip to Chanak.

The first attempt failed, I was discovered, forced to dive, turn around and give up the attempt. But I succeeded in the second attempt. However, I was almost discovered when I arrived near the south-east corner of the Imbros Island, close to land in order to cover the open area surfaced up to Cape Helles. The northernmost destroyer of the outpost line passed me at about 150m in the dark night without seeing me.

I considered in a split second. Dive? Then the enemy hears the roar of the air swirling out of the diving tanks. Then I am hunted with weak batteries...

"Heavens!" Whispered my lieutenant.

Light signals blinked from land. Blinding such that one had to close one's eyes, the world was shrouded in the deepest darkness when they were extinguished. I owed my life to these signals.

The destroyer had so much to do with the reading that he did not notice me shrouded by the black coast – the signal station was very bright.

I drove on peacefully and was very much relieved when the enemy destroyer disappeared to the south. Shortly before dawn I dived near Cape Helles and unseen and undisturbed managed to get to Chanak with completely spent batteries. Fortunately arriving in Constantinople, I replaced the damaged battery cells with new ones and then ran out again for further cruises.

Steamer Carthage

Carthage going

Going

Gone

VIII
Carefree Days in Pola

A poisoned "love gift" from neutral ships?

In November 1915 I sailed from Constantinople to transfer my main base of operations to the Austrian naval port of Pola.

It was also absolutely necessary for U 21 to be once again thoroughly overhauled. Insufficient spare parts could be brought down to Constantinople and finally I had to obtain new torpedoes again.

However before I could go to Austria I had once again to use Turkish assistance: an oil cooler had leaked on board and as a result I had a very high consumption of lubricating oil, I could calculate the day when my lubricating oil was all spent and my engines would run hot.

I therefore decided to proceed to the Gulf of Adalia (on the southern coast of Asia Minor) and try to obtain suitable oil there. It proved difficult to obtain oil down there since they had hardly ever seen an engine.

Ultimately, my Chief Engineer had the good idea to try olive oil, which was very cheap and available in large quantities. What did I want to do? I said "yes". Then we continued with olive oil as lubricating oil.

Incidentally, the Turkish commander was very accommodating regarding the procurement of oil and when we were leaving donated two freshly slaughtered chickens for each crew member which we only had to pluck. The donation was accepted with many thanks because after the eternal monotony of canned food, any change was welcome even if it was just chicken soup.

In Pola as in Cattaro I was welcomed with the greatest warmth and kindness by our Austrian allies. Every wish, even barely expressed, was immediately fulfilled whether it was crew accommodations or the provision of supplies, or assistance in repairing the boat in the Pola Arsenal or on the workshop ship "Gäa"[39] in Cattaro.

The Austrians also provided all kinds of variety. If time allowed, we made short and long trips by automobile from Cattaro to the mountains at the Montenegrin front and we received the same warm welcome everywhere. I will also never forget the days we spent with Austrian submarine comrades on the island of Brioni off Pola, a small paradise. In this way our Austrian comrades made sure during the time in harbor between two long-distance voyages we were able to take our minds away from other matters and temporarily forget the severity of submarine warfare.

The Austrian fleet commander, Admiral Haus[40], was also very interested in our submarines and received personal reports every occasion we arrived.

Finally, I want to tell you a little episode I experienced in the Mediterranean demonstrating the means by which war was waged against submarines.

I stopped a Spanish steamer to examine it. While I was checking the papers and my demolition team was studying the steamer, the steamer suddenly made a signal.

I read in amazement: "Do you need provisions?"

I was astonished at this kindness. Nice, very nice ... just ... somehow the suspicion suddenly stirred. If ... you're a fool, I thought to myself. But the distrust could not be dispelled and remained. Since I had enough provisions on board, I politely declined the offer with thanks.

---

39 SMS "Gäa" was a torpedoboat flotilla auxiliary vessel of the Austro-Hungarian Navy placed at the disposal of German submarines as a depot and accommodation ship.

40 Anton Haus, later Grand Admiral until his death early in February 1917.

Several months later I found out at home that another of our submarines had also been supplied by a Spanish steamer; the entire crew had fallen ill from this food. Would a truly neutral steamer have lent itself to provide German submarines with poisoned provisions?[41]

U21 in Austro-Hungarian waters acting as U36 under the Austrian flag as show in the photo.

[41] Hersing sank no Spanish flag vessels; the Spanish ship's papers must have been in order in terms of not carrying contraband.

IX.
One man saved - from the entire crew of a Battleship!

Sinking of the French armored cruiser "Admiral Charner"

At the beginning of February 1916, I cruised in the eastern Mediterranean to wage war on trade off the Suez Canal. One evening when listening to enemy F.T.[42] traffic I realized enemy warships had to be nearby. Since I knew the French often used warships to fire at Turkish cities on the Syrian coast, I drive north during the night to be in position in front of Beirut's port at dawn the next day and if I as lucky a French warship should arrive there to bombard the port.[43]

It was early morning on February 8th and I was correct again in my assumption. I climbed down for a brief spell from the tower on which I had been standing all night to use the map and determine my exact ship's location. No sooner had I reached below deck and spread out the map than a message was passed to me through the boat that a large cloud of smoke was in sight from the North.

Immediately back into the tower ...

It was just beginning to dawn, I had to get my eyes accustomed to the darkness.

Out of this darkness came a fat cloud of smoke ... probably a powerful ship ...

---

[42] F.T.= Funken Telegraph or wireless.

[43] Hersing omits mentioning defects to his engines limited surface speed to 12 knots, he decided to deviate from his mission of disrupting merchant traffic off Suez and went in search of warships along the coasts of Syria.

He became more apparent when I took a longer look: a grey wall, the background... it had to be a big warship!

"Quick dive!"

The air whirled out of the diving tanks filling with water.

I turned towards the enemy ...

Less than five minutes later I could see him clearly. It was the French armored cruiser "Admiral Charner" ...

He was at high speed, bow waves foaming ...

But he wasn't steering towards me, he was zigzagging ...

It was difficult to aim properly ...

Maybe he would still come straight to me. I waited and waited ... but the gap began to increase, rather than decrease ...

And always in a zigzag course ...

Despite the zigzag course, I came close to the Frenchman at 2000 m ...

Then he turned away. If I did not shoot immediately, he would escape ...

I waited until he changed course again, he nicely showed me the broadside ...

"First torpedo ready ..."

A light pressure with the finger ... the boat shook, the torpedo is out ...

Now I waited ... waited ... until the torpedo ran its two kilometers ...

Now, if my calculations were not incorrect, the shot had to hit ... if I had correctly estimated the speed of the armored ship and also had led him enough to reach him, then it must...

Aaaah ... smoke and a water fountain rose sky-high ...

Such a shower ... I hit right under the bridge ...

There had to be a hole for three railroad trains to enter side by side ...

What followed literally happened in a minute and a half: the ship capsized to starboard ...

I look at the clock ... not a minute has passed, the entire hull of the ship is already submerged, only the masts and funnels protrude from the sea as if in supplication ...

Suddenly the ship turns over ... revealing a red keel ... the screws visible ...

Is it haunted? The screws race on ... the dead ship's machinery is not turned off ...

I have never been under the impression a ship has a soul... as if it were not a machine, but a living being with reason and strength, suffering and agony ...

A terrible detonation ... the boilers exploded. It was as if the ship had risen for a moment ...

Then there was only a black cloud of smoke over the spot ...

At that moment Lebanon was shining in the flashing white of its peaks ...

One man[44] ...a single one of the entire crew, horror in his eyes, drifting, hopeless in the sea.

He was saved.[45]

---

44 Quartermaster Joseph-Marie Cariou. Recently assigned to the ship, onboard for 15 days prior to the sinking. His testimony: "On Tuesday morning February 8, we were near Beirut, about 10 to 15 miles offshore. I was on the aft deck. The weather was fair enough, not very cold, very little wind, ... Having heard a thud with a strong shaking of the ship, we immediately had the impression that it was a torpedo, because the night before, while setting sail from the Isle of Rouad, a submarine was reported. The boat nosed down a little, and capsized almost immediately. It was then that I wanted to take off my jacket. I didn't have time, I was thrown onto the guardrails and I sank with it. When I came back to the surface, I found myself next to a small piece of wreckage, a demolished hen cage , and I stayed on it for about an hour...."

45 The German Official History on the Submarine War notes:

Reading his logbook the Naval Staff in Berlin gave the following instruction to the Pola Submarine Flotilla:

"The logbooks of U 38 and U 21 show that commanders deviate from their orders and spend a very large part of their time on operations against warships. Commanders are reminded the most effective way to bring down England, our main enemy, is to increase as drastically as possible the shortage of goods from

French armored cruiser Amiral Charner

---

which she suffers. As satisfactory as the destruction of warships is, it is for the moment a much less important means of action than the destruction of merchant ships. This does not mean, however, commanders whose results are otherwise fully appreciated should systematically refrain from taking advantage of circumstances, when they are favorable, to destroy warships. They may even be authorized for this purpose to depart briefly from their operational orders, but war on trade operations should not be neglected in favor of operations against warships ... ".

X
The Submarine, the "Girl for Everything"

As a transport ship to Africa for the Senussi

Due to dense enemy patrol and the blockade activity, it was almost impossible for Germany to undertake long voyages by surface ships to support allies by sea. The submarine had now also to serve this purpose and was used as a gofer for everything. In the summer of 1916 I was ordered to carry arms and ammunition with my submarine from Pola to North Africa to the Senussi.

The Senussi were in battle with the Italians at Tripoli, south of Bengazi and were now to be supported by us in order to tie up more Italian forces in North Africa.[46] I packed my boat full of carbines and rifle ammunition, as much as the boat could carry; and proceeded with four Turkish officers through the Adriatic and the Mediterranean to the African coast.[47] Through the mediation of the Turks, a rendezvous point was arranged with the Senussi, where they could receive the weapons.

However, this meeting point was very difficult to find because the entire coast appeared the same everywhere for miles: high dirty yellow sand dunes with some undergrowth on them, otherwise no conspicuous point to head towards. Eventually, I was shown from land a white flag on a long pole, which would serve as a recognition sign. I drove carefully towards the coast,

[46] The Senussi are a Muslim political-religious order whose opposition to French and Italian colonial expansion predated the Great War. The Central Powers hoped to incite further operations to force additional troop deployments of British (Sudan), French (Morocco), and Italian (Libya) forces.

[47] This section corresponds to events transpiring in April 1916.

which was little surveyed[48], dropped my small dinghy, and one of the Turkish officers went ashore as an interpreter.

He soon came back on board with a Senussi, dressed very picturesquely, in a dirty white burnus[49], around him hung several cartridge belts strapped on, a large dagger knife in his belt and a man-sized shotgun in his hand.

This Senussi would serve as my guide to a Senussian camp where I would hand over the weapons. The next night I drove to this point and was first received very suspiciously by blacks who had dug trenches ashore and who were ready to attack if I attempted anything against them. However, landing at this point was not possible due to the high coral reefs lying off the coast and protruding from the water. I arranged another meeting point with the black men further south where I could get through a small passage behind the coral reefs to try to get the heavy boxes of weapons and ammunition ashore.

At this point, the remains of an old Turkish barracks were still on land, which had been razed and destroyed during the Tripolite war.[50] I had arrived successfully and anchored behind the coral reefs; but disembarking the ammunition was not so easy, as my dinghy was too small to carry even one of the heavy boxes. I had to think of another solution. The Senussi arrived the next morning with a large camel caravan to carry their weapons away. With the help of these Senussi we made a large raft out of the remains of the wooden barrack gates on land which had enough carrying capacity to hold a box and two men comfortably. We now connected the boat to the land by a strong hemp hawser and via this connection the raft was slowly brought back and forth and in this time consuming manner the whole load was brought ashore within three days.

---

[48] Reference is to hydrographic survey.

[49] The Arabic word for the white attire & headdress worn in Arabic countries. Commonly expressed as burnoose in English.

[50] The 1911-1912 Italo-Turkish War. Turkey conceded its rights over Tripoli and Cyrenaica to Italy.

But the landing had not taken place completely undisturbed. One morning, the Italians appeared with patrol vessels to search the coast. I was able to get just outside the coral reefs and then dive quickly before the enemy saw me. After a few hours the vessels left again because the Senussi hide with extreme swiftness the ammunition already landed behind the nearest sand dune and the Italians could not discover anything unusual.

On the second night, a heavy swell suddenly arose and the boat rocked hard in the ground swell and constantly pulled at the anchor chain. I didn't dare to run between the narrow coral reef passage because of the towering surf as I could easily have been thrown onto a reef. A disaster suddenly appeared when my anchor chain broke, nearly tossing the boat onto the beach. Only by immediately starting the engines was I able to keep the boat from running aground. But the anchor and part of the chain were lost on the bottom. The disembarkation of the weapons could continue after the weather calmed down. Since I no longer had an anchor, I constructed an emergency anchor by tying together several heavy lead weights that were on board to trim the boat and attached them to the rest of the anchor chain. After everything had landed and with much thanks from the Senussi, I went back to sea for the other part of the operation, waging war on trade in the Mediterranean.[51]

[51] Various submarines executed supply and liaison runs to North Africa during the war until the effort was terminated as the time consuming voyages detached the boats from focusing on their main purpose of waging war on seaborne trade.

Typical supply mission to the Senussi.
This photo is from an offloading from UC20

XI.
The Unrestricted submarine war was two years late

Bethmann–Hollweg's Role

In 1909, all of the world's seafaring nations had gathered around a green table in London and agreed in international agreements what to do in the event of war with enemy and neutral merchant ships and their cargo, which were all privately owned. These international agreements were called the "London Convention of 1909". And what was stipulated in it was termed "Prize Law".

According to which each steamer had to be stopped by a warship and then have its cargo examined which was divided into free goods and contraband goods.

This contraband was again divided into absolute and relative contraband.

Then it was determined exactly under which circumstances a steamer, both hostile and neutral, could be taken or sunk.

The legality of a warship's actions was subsequently decided by a prize court in the port.

Submarines also had to follow these rules, of course, although shortly after the start of the war England had her Naval Minister declare England did not consider herself bound by these agreements of the 1909 London Convention.[52] The action followed: England commenced a hunger blockade against Germany.

---

[52] None of the signatory nations ratified the agreement and it did not come into force during the Great War.

This submarine trade war went well for a while until the enemy began arming his steamers resist their investigation contrary to any international law. At the insistence of the Front[53], then Chancellor Bethmann-Hollweg finally decided to declare unrestricted submarine trade warfare for the first time in February 1915, i.e. we were no longer bound by the terms of the 1909 London Convention and would treat the crews of armed steamers as pirates, as it was in international law.[54]

No sooner had our declaration been announced than the enemy press shouted about our brutal warfare which showed the world how we were similar to Huns and Barbarians. As a result, Mr. Bethmann-Hollweg was afraid and after barely 14 days the unrestricted submarine trade war was called off again, and only the 1909 London Convention agreements had to be followed again.

According to Bethmann-Hollweg remarks we should conduct the war "humanely" and not irritate the enemy any more. As if one could provoke the enemy any further in war where life and death are at stake. Severe punishments were threatened to submarine commanders who from now on did not abide by the London Convention. The submarine trade war took place in the following manner:

Before I could examine a steamer a long or short artillery battle resulted each time until the enemy struck his flag and got into his boats to come alongside. It was very interesting to observe the individual races. The Romans,i.e. French and Italians raised their white flags when there was only one hit on the steamer and were glad to have escaped certain death, while the Englishman in most cases defended his steamer with extreme tenacity. I myself once shot at a steamer for about three hours and the Englishman only gave in reluctantly and raised the white flag because he received a hit from me in the main steam line. He

---

[53] Meaning naval officers involved in the submarine war at various level of command, from boat commanders to senior levels.

[54] A notable example being Capt. Charles Fryatt who was shot after capture as an illegal combatant for attacking a German submarine with his merchant vessel.

had several dead on board and also wounded in his boats, which I even bandaged for him as he had no bandages on board.[55] And all of this was not undertaken by an active naval crew, but by a civilian crew that should have been treated as a franc-tireurs[56] under international law and should have been hanged under international maritime law.

And we had to wage war humanely!!

When a steamer was overpowered the crew abandoned ship. If the steamer was deserted and there was time and opportunity, you sometimes went alongside before it completely sank and checked whether there was something on board worth taking. The first area examined was the machinery storage compartments where steamers mostly had a lot of copper and brass and rubber seals in reserve. These were all so-called rare metals, which were very scarce at home, which was completely closed off from the outside world and we could put to very good use in a submarine, especially at the end of the war even the most important pipelines and fittings in submarines were already so-called war substitutes, and we were happy when we got top grade material again.

What we could not use ourselves was handed over to the shipyards at home so that other submarines could benefit. The provisions load was also gladly examined, but what was searched for first was not alcohol, which some readers might want to assume, (Because most landlubbers live with the idea that a sailor always has a thirsty throat because he sees so much water all day) no, what was searched for first of all was fresh bread. We could never get enough of it. It was impossible to bake on a submarine and the fresh bread we took with us from harbor would last at most 5 days, then the smell of oil, etc. in the submarine made it completely inedible. We only lived on so-

[55] The vessel in this engagement has not been identified as of this writing. Hersing is reputed to have assisted the survivors of the British steamer "Belle of France" in February 1916.

[56] Literal French "Free Shooters". Civilians who took up arms contrary to the rules of war. Unlawful combatants in contemporary usage.

called hard bread or ship's cracker, which was carried onboard soldered in large tins. To crush this hard bread you had to use sturdy tools, preferably a large sledgehammer. If anyone was lucky enough to find maggots in their bread, he was glad he had none fat-free day. The rest of the food on a submarine consisted only of canned food that was electrically heated during the long of 4-6 weeks cruises. These meals were always repeated in a uniformed defined sequence. There were no special delicacies on a submarine and if we ever had the opportunity to get fresh bread or perhaps ham, sausages and sides of bacon from a steamer, we were happy to do so.

This submarine trade war now continued despite the warnings from the Front that we cannot achieve our goals in this manner. When a submarine shot at a steamer somewhere on the seas, the steamer naturally reported its distress to all areas by spark telegraphy, in order to summon the enemy patrols to help, who then attempted to chase or destroy the submarine. It also alerted all other steamers in the area so they could immediately flee before the submarine had finished the first steamship. Mr. Bethmann-Hollweg was finally also convinced and forced to declare an unrestricted submarine war.

XII.
Depth charges, nets and listening devices against us

The English intensive defensive struggle

When we started our submarine war on trade at the beginning of the war the Englishman immediately recognized this type of warfare went after his lifeblood. He therefore spared no means and expense to expand his countermeasures against submarines. At the war's commencement the submarine was a completely new weapon that had never appeared in world history. No one could know what role the submarine would play in a future war. Therefore countermeasures against submarines had not yet been developed. But the Englishman immediately formed a large ministry with experts who were only concerned with devising and testing defensive measures against submarines. He felt that his large number of torpedo boats and destroyers were not sufficient to successfully fight the submarines, so he immediately built a large number of guard vessels, patrol boats and submarine hunters. The latter were small, shallow draft, high-speed vessels mainly equipped with depth charges on board as main armament alongside their light artillery.[57] These depth charges were large explosive devices that were simply cast overboard near or over the submarine; however, they did not detonate when they hit the water, but were set with an adjustable detonator to a certain water depth, at 10, 15, 20 m, even deeper, in order to explode as close as possible to the submarine.

[57] Most likely a reference to P-boats.

The Englishman was very generous with these depth charges. When he had determined the approximate position of a submarine, he drove over the boat at high speed and cast 6-10 units one behind the other. Then the next one came and it went perpendicularly over the submarine and so on. A submerged submarine was really not in an enviable position during this time. First of all, the detonations in the submarine boomed like in a sounding board, everything not riveted and nailed down was tossed about. The shock splintered electric lamps , seals flew out and water came into the boat, etc.

It was most dangerous when an oil bunker leaked due to a depth charge, because oil immediately came to the water's surface and the enemy could pinpoint the submarine's location and course in this manner and continue with redoubled effort his attacks on the submarine using depth charges.

Furthermore, the Englishman had strengthened his port entrances not only by mines, but also by submarine nets, some of them he also dragged in shallow water between two steamers in order to catch a submarine in this way. The submarine, with its hydroplanes and screws, could be trapped in the net and thus forced to surface. A surfaced submarine is always easy prey for a surface warship. A direct hit in the pressure hull renders the boat unable to dive and makes it easy to sink. A submarine's main defensive weapon is its ability to dive and place the protective water surface between itself and the enemy.

Furthermore, in order to prevent enemy submarines from approaching its coasts, the Englishman had surrounded his entire island with mine fields, which were not designed for a depth of 3 - 4 m, as was the case with ordinary mine fields, but for 10-20 m to destroy submarines who went deeper.

Finally, he had surrounded his entire coastline with a system of underwater eavesdropping receivers, small membranes he anchored underwater out at sea and were connected by cable to a central location on land. If a German submarine approached the coast unseen as possible underwater it immediately betrayed itself with its propeller noise. The central office on land located the position of the submarine precisely through various membranes and signaled the location to the guard and patrol vessels, who of course immediately ran out at high speed and in large numbers to destroy the submarine with depth charges.

When we were first attacked by the enemy in this somewhat inexplicable way, without being seen by the enemy, we first believed it spooky until we finally figured it out. The Englishman even equipped all his warships and steamers with underwater sound receivers in order to be able to detect the presence of a submerged submarine near the ship and determine the direction in which the boat was located, so he could escape the submarine using high speed. He also tried to completely block the passage through Dover-Calais to the Channel with nets, deep mines and a multitude of guard vehicles. At night he even had a large number of steamers in the Strait, which lit the Strait with large, burning magnesium fires on their decks at night to make it impossible for submarines to get through surfaced in the dark.

In doing so, he wanted to prevent his communications between England and France, where he had to transport a large amount of war material, to be disturbed by German submarines. Despite all these measures he was unable to keep German submarines away from thc high seas. That's why he resorted to another means. He wanted to make it impossible for us to leave our home ports. He tried to block the ports on the Flanders coast, Ostend and Zeebrugge, in a boldly designed and with extraordinary courage and energy attempt using blocking cruisers, which he sank in the port entrance. In Ostend he sank the cruisers next to the port entrance, in Zeebrugge his cruiser was able to reach between the moles but was unable to block the entire exit.

It was impossible for the Englishman to get to our estuaries in the German Bight in the North Sea. That's why he systematically contaminated the entire German North Sea Bight with mines to prevent submarines from exiting through the North Sea. The way through Sound and Belt from the Baltic Sea around Skagen was much easier to guard. The English planted about 9 million mines in the North Sea during the war.

But despite all of this, thanks to the self-sacrificing work of our mine detection and escort flotillas, we always managed to keep several mine-free routes clear through the North Sea. The Englishman had to see the impossibility of keeping our submarines away from the high seas.

Therefore he decided on the latest measure by stating steamers should not continue to travel alone. However, since he did not have enough patrol boats and destroyers to surround every steamer with the necessary security, he resorted to the tried and tested means of a convoy. In the harbor or out at sea, he assembled a large number of steamers, which traveled in tight grouping on zigzag courses and surrounded all of these steamers with a huge system of security vessels. Far ahead of the convoy steamed a cruiser which hoisted a balloon in calm weather to have a larger field of vision. As soon as this cruiser spotted a submarine, he signaled the presence to the convoy so it could turn around, or more specifically evade the submarine. Behind it was a large, widespread veil of modern destroyers, which would shroud the convoy from sight of a submarine with a lot of smoke and high speed. In that event, in front of the steamers and on both sides, a number of destroyers went back and forth at high speed and zigzag courses at a distance of about 300 to 2000 m, the most favorable firing distance for a submarine. If a submarine reached the convoy unseen and showed its periscope shortly before firing it was bound to be seen by destroyers moving back and forth at close range. Behind the convoy sailed a number of escort vessels who then, if a submarine managed to get off a shot, (and it had betrayed its location in this manner for example), destroyed it with depth charges.

XIII.

Racing a Torpedo

The scariest race of my life

In the previous chapter I briefly mentioned English countermeasures against submarines. The Englishman knew he was threatened by submarine warfare and therefore led the fight with all his energy and tenacity. He even built some vessels, the so-called submarine traps, intended to destroy submarines. These were normal steamers of about 4000 tons which had nothing more than cork and empty herring barrels as cargo so the steamers would float due to the cargo in event of being torpedoed by a submarine. On the upper deck he had a hidden 15-centimeter battery. This submarine trap now lured the submarine closer due to her behavior, for the most part still flying under a neutral flag, in order to be examined. Woe to the submarine which was duped by the trap's behavior and came too close. The Englishman immediately folded down the flimsy bulwark and showered the submarine up close with a volley of his 15-centimeter guns. The following incident proves how excellent English artillery and range measuring equipment was on these submarine traps:

I met a steamer, not too big, not too small. Nothing else caught my eye, he didn't stick out of the water like other submarine traps which had loaded cork or were empty ...

I called to him but he resisted. He fought back with a miserable 5-centimeter gun that spat on me. He did not zigzag but continued quietly on his course as if I were not there. I came closer. As soon as I was at 4000 m I turned broadside and give my two guns permission to fire.

I observed my impacts: they landed perfectly. The next ones had to be amidships ...

Something surprising happens (if one can still speak of surprises in this fight). The little steamer loses his nerve, folds down his bulwark, I see two powerful guns, at least fifteen ...

A whimpering is already on the way.

I hear the parabola singing through the sky: we burst into the submarine as quickly as we can.

But the whiz comes closer and closer and does not intend to move left or right: ssssSSS ...

God, that has to be a direct hit ... The shells detonate so close a poisonous smoke drags across my face in ugly swaths ... then it suddenly grows dark around me.

Then another heavy blow: the boat has been hit ... I can see at the moment the net deflector has broken off.

Dive? Impossible, if a shot landed before I got under water, the boat would be destroyed ...

"Flank speed ahead!"

Beneath thick smoke exhaust I steer away from the ship. Using a zigzag course. I do not think it is going to end badly. His salvos are close to the boat. I move to 8000, 10000, 15000: the salvos are still close to the boat.

It was a pure coincidence that I wasn't hit. Also, if I had come closer to the steamer at the beginning of the battle: I would have been destroyed like so many of my comrades. [58]

On this occasion I must also mention the well-known Baralong Case, where the Englishman, after destroying the German submarine with artillery fire also shot with machine guns at the defenseless German submarine crew swimming in the water. Unfortunately, this case has not remained isolated. Otherwise the Englishman was generally a very decent opponent.

---

[58] Incident above relates to Hersing's encounter with Q.ship HMS "Century" on October, 22, 1916. The Q.ship was armed with two 4"(10.2 cm) artillery pieces.

The Brit also tried to fight our German submarines with his own submarines. At certain prominent points where he knew German submarines were lurking for enemy steamers he was there as well with his submarines and it was now a question of which of the two submarines arrived and had seen first while submerged. It happened to me once on the southwest tip of Ireland, at Cape Fastnet head, where I was lying in wait to intercept steamers arriving from America and headed for this Cape as the first landmark.

I traveled surfaced in the morning and watched carefully to see if there was an enemy submarine in the area.

Nothing. Absolutely nothing. Nobody was there ... no surface vessel, no surfaced submarine to see and no periscope. Nothing.

I calmly proceeded up and down. Nothing.

I clearly remember having just checked my watch: 9 o'clock. I took the scope and swept the area again: oho ... a periscope abeam! I immediately understood: an enemy!

I no longer had time to think: I had to act immediately. Veer away at the maximum speed, show our opponent the stern ... !!

But our opponent was also quick. I saw four air bubbles coming at once, rolling up towards me, so to speak, at high speed.

The first torpedo suddenly curved away... it was past ...

"Flank Speed ahead!"

Then the other two torpedoes barely passed my stern by a meter ...

But a fourth torpedo was approaching. I could no longer avoid it ... it came straight at me ... as quickly as possible I steered my boat exactly parallel to the torpedo track ...

The torpedo raced closer, closer. We were staring at the scary projectile which could be seen in its entirety in the bright sun.. if he just doesn't... Devil... He ran, yes, he ran exactly one meter next to us... was past the stern... We held our breath in the tower. What if he doesn't turn? What?

The further he overtook the boat, the closer he got to the ship's side, because the boat was thicker in the middle!

Suddenly a thought: I climbed out of the tower: try to keep off the torpedo with my boots ...

There ... the torpedo jumps out of the water like a pike ... it sinks headfirst ... ah, that was a relief!

And spontaneously we sent the British over there three happy "Hurrah!", throwing our hats in the air.

The miracle, of course, hadn't been accomplished by my boot: the torpedo had run off ...[59]

[59] The failed British submarine attack occurred on April 29, 1917.

XIV.
Towards the bottom of the sea at frantic speed

The boat is slowly crushed - flooding -All seals burst

With what bravery, tenacity and intelligence the English had taken up the fight against German submarines and what means they used to destroy the German submarine plague, as they put it, I have indicated in the previous two chapters. I would now like to briefly tell you about some of the experiences I had during the unrestricted submarine war.

It was in 1917 when I attacked a well-escorted convoy. In order to escape depth charges as quickly as possible after the torpedo shot, I performed the following maneuver with my boat: As soon as the torpedo left the boat I gave the command in the Control Room: "Flood!" "Go to forty meters!" "Flank Speed ahead both motors!" At the command "Flood!" The bottom valve was opened to allow water around the pressure hull and make the boat heavier so it went deep quicker to avoid depth charges. I went flank speed ahead to get the boat under the convoy by the quickest possible route because the enemy would not look for me there and could not throw depth charges on my head. [60]

Until now the maneuver had always been satisfactory but this time it was to be different. During my attack I was discovered by the flank escort and just before releasing my torpedo I heard the sound of screws and the roar of the enemy destroyer behind me. Nevertheless, I did not let go of my attack and after the torpedo was out I immediately gave the command: "Flood!" "Go to forty meters!"

If only no depth charges now ... I had been too impudent.

---

[60] Hersing elected this maneuver after being hunted for several hours following a convoy attack in April 1917.

A terrible blow ... Splinters hail through the boat, all lamps go out, extinguished: it is pitch dark and we are half dead: hits ...

My flashlight lights up. I feel pale ... either the depth gauge is damaged ... no, no, it is not damaged, it is very much active: we are sinking like a stone !!!

I immediately receive a message from the Control Room: "The bottom valve is stuck in the open position ... it just can't be closed!"

"All pumps instead ...!"

The seconds become hours ...

"The water has been pumped out," is the report. But what good is it: the ship has negative buoyancy ... we are sinking!

The manometer has gone crazy: 60 ... 63 ... 70 meters!

I order in Control Room: "Attempt by all means to close the bottom valve, I will catch the boat with compressed air from the tower."

Compressed air is first applied to the middle ballast tanks.

Initially only two, then four, and finally I blow 150 Kg of compressed air into the six ballast tanks on each side ...

But the boat continues deeper!

Depth charges hail down.

The manometer shows 85 meters ...

Ninety!

Then it finally stops. Because the pressure gauge pointer cannot go any further: the scale is no longer sufficient ...

A fine drizzle from all sides: the rivets are no longer tight!

Seals pop out ...

What's this? The horizontal iron plates on which we stand bend up like sheet metal: the boat is being crushed !!!

We had to be 100 m deep. The pressure on 1 sqm was 100,000 kg! And under my keel the sea was still 3000 m deep at this point !!! ...

Compressed air! Compressed air ... I give more and more ... it is now or never! Compressed air! Compressed air!!! Press ...

There: the manometer is stirring again! Shouldn't we be drowning?

Suddenly from the Control Room: "The bottom valve is closed!"

Aaaaaaahhhh !!!!

"Establish normal trim, boat at forty meters!"

I immediately opened the large top valves to let air out. Because the boat would have jumped like a rubber ball on the water's surface with all this compressed air inside the hull ...

Saved? There was no certainty: by expelling the compressed air, the enemy now knew exactly where we were, depth charges would surely now come ...

But strangely they did not come. Dead silence. Well?

We had the peace and quiet to get our boat back onto 40 m and repair the damage.

Such situations can only be mastered with a well-trained crew with a strong sense of responsibility and above all mutual trust from the commander to the youngest fireman and vice versa. Only when there is this mutual trust and feeling of comradeship, meaning: "all for one, one for all", can one be up for such situations. The commander up in the tower must be kept informed of all operations within the boat and if for any reason an order cannot be carried out, the commander must be notified immediately so that he can take countermeasures to rescue the boat.

When I returned to home port about three weeks later the riddle behind why the enemy suddenly stopped throwing depth charges was solved. I submitted my war diary and report regarding the events of the voyage and presently the English war report has been compared and the latter reported a German submarine attacked a convoy and destroyed a steamer on the same day and in the same place where this incident happened to me. The submarine had been destroyed in the process: that would be me! When I let the compressed air out of the ballast tanks and air came to the surface with a big gush, the enemy assumed the boat had broken apart and the last air had now come to the surface.

Another time I was lying northwest of Ireland in a heavy Atlantic sea and wanted to go a little further to the coast to find possible attack opportunities there. There was no vessel to be seen anywhere in the poor weather, but the Englishman, although I was submerged, caught me through his listening apparatus and a flotilla of destroyers was already chasing me to work me over with their depth charges. Trying to escape to deeper water, I struck the aft hydroplane against a rock such that I broke the yoke of the hydroplane. The rudder could no longer be moved and there was no longer any thought of submerged travel. I had no choice but to drop the boat to the bottom and wait and see what would happen next.[61]

After I had turned off all the machinery, the enemy could no longer determine my presence with his listening devices and gradually abandoned the pursuit.

I thought about what to do. Repairing the rudder was impossible; I could still dive, but only to a point to lay the boat on the bottom. So I had no choice but to try to get home surfaced through the enemy security lines. After dark, I dislodged the boat from the bottom and came to the surface. Here the heavy Atlantic seas became an unexpected benefit for me: In the poor weather there was no vessel in sight and I traveled surfaced largely west and north, through the security line between the Shetland Islands and Faroe Islands at night and tried to reach the Norwegian coast while surfaced. Continuing to Wilhelmshaven through the North Sea was completely hopeless because of the many enemy minefields planted there which of course the English guarded. So I drove south under the coast of Norway and arranged my voyage to positioned in the evening offt Skagen, the northern tip of Jutland, to pass through the Kattegat and the Sound to the Baltic Sea at night. I arrived at Skagen after dark

[61] This event corresponds to February 18, 1918.

and continued driving south through the night. Here I passed surfaced an English submarine on duty, also surfaced, which lay there to make sure no enemy could get around Skagen unseen. It was an advantage for me that I arrived in front of heavy seas on a dark night while the Brit was driving against the sea and had a hard time seeing ahead because of the spray.

Towards morning I arrived before the entrance of the Sound and since I was not allowed to be seen in the neutral waters I put the boat on the bottom on the side of the entrance in order to surface in the evening. After dark, I surfaced and drove south through the Sound at high speed. It was very difficult to pass the narrowest part of Helsingborg, where a lot of brightly lit pleasure boats and ferries bobbed up and down. Naturally I tried to sneak through all the vessels completely dimmed but I was seen by a small Danish torpedo boat on watch while I was driving past the brightly lit promenade. He immediately turned towards me to find out what kind of Flying Dutchman was driving around here at night, completely darkened. After chasing me for about an hour (he could easily see my bright wake in the water's luminescence), he turned on his spotlights to see who I was. However, when I showed him the German battle flag and called over to him to switch off his searchlights, because otherwise he would blind me, he was satisfied and let me continue unmolested. I passed this night through the Sound near Malmö, under the Swedish coast and into the Baltic Sea, where I reached my home port Wilhelmshaven safe and sound via Kiel and the Kaiser Wilhelm Canal, albeit with a badly damaged boat.

XV.
"Have you sunk six Dutchmen?"

A Question of Authority and its Consequences

When unrestricted submarine warfare was finally declared in 1917, I was ordered to return to Germany from the Mediterranean to fight the intensified submarine war from home. In February I started my return trip to Wilhelmshaven from Cattaro. A chain of command question now arose from the higher authorities as to whom I should report to during the whole trip. In the Mediterranean I was subordinate to the Mediterranean Submarine Flotilla, then came through the area of the Flanders Submarine Flotilla and finally into the command area of the Submarine Commander in Wilhelmshaven. In order to make things as simple as possible, the Naval Staff in Berlin ordered U21 to be subordinate to the Naval Staff in Berlin from the moment they left Cattaro until they arrived in Wilhelmshaven and only to receive orders from them. Therefore I received all my orders, especially regarding the commencement unrestricted submarine warfare in writing directly from Berlin.

The consequences of this highly embarrassing conflict of authority would be quite large. I did not stay long in the Mediterranean and only wished to accomplish my principal work in the blockade zone in the Atlantic. I managed to get cleanly through the Strait of Gibraltar surfaced at night despite the heavy guard. In the Atlantic, I now started submarine trade warfare and gradually came to the exit of the English Channel.

Here I discovered a large cloud of smoke emerging from the Channel, which developed into six Dutch steamers. Despite the intensified submarine warfare, these steamers huddled together through the restricted zone. Since it seemed a little strange to me

I read through all the Naval Staff orders again and found it said not to be lenient after the beginning of unrestricted submarine warfare, especially against neutrals.

I then stopped the steamers with gunfire. They immediately stopped and came alongside me in their boats. When asked if they did not know unrestricted submarine war had been declared and they should actually be sunk here in the restricted area without warning, they replied it was well known to them, however they had conditional security from the British Admiralty in London that they might depart from Plymouth to the Atlantic Ocean around the north of Britain and over to Norway to Holland."

I gestured to them that I was tremendously sorry but the English Admiralty in London could not dictate to me what to do; would they not have been given security by Berlin?

I proceeded to sink the steamers when they answered no. I took one of the Dutch boats in tow — they happened to be old acquaintances, for I had previously stopped their steamer in the Mediterranean about a year ago — and now went with this boat from one steamer to another and my demolition crew could board using this boat to sink the steamers. The other boats returned to the English coast.[62]

An English destroyer appeared at dawn when I had just finished with the last steamer, he had apparently gradually noticed there was a German submarine at work here on the coast. I paid the Dutch crew bread and a bottle of cognac and dived to escape the Englishman's eyes.[63]

---

[62] The "Bandoeng", "Eemland", "Gaasterland", "Zaandijk" and "Menado" were blown up with explosive charges. "Menado" was able to be towed to safety. The "Jacatra" and the "Noordwijk" were torpedoed. The ships were loaded with grain.

[63] Dutch eyewitness reported Hersing was agitated on occasion (particularly when presented with the safe conduct orders by Dutch captains) and was seen to be keeping a watchful eye for British patrols throughout the attack. The Dutch vessels had departed Falmouth and were in waters certain to be regularly patrolled.

When I arrived home I learned there was something special regarding these Dutch steamers. I had finally come north around England through the North Sea to Helgoland to continue the next morning to Wilhelmshaven. As soon as I landed in Helgoland, where I was welcomed with music, the Island Commander greeted me with the question: "Have you sunk six Dutchmen?" When I said yes, he just smiled and said: "You'll find out the rest in Wilhelmshaven." I couldn't get anything else out of him. Afterwards, when I was sitting with other submarine comrades, they told me every incoming submarine was received with this question, and I gradually got to the bottom of it.

The Foreign Office in Berlin had once again compromised and assured the Dutch these six steamers would not be sunk by German submarines although unrestricted submarine warfare was already in effect. The Naval Staff instructed the Submarine Command in Flanders and the North Sea to ensure all submarines that were departing and at sea were contacted by F.T. and notified of the agreement every night for eight days consecutively. Only I alone did not received the radio message because I had another cipher key that only the Naval Staff and I knew. I could not decipher the radio messages made by the Flanders Submarine Flanders or the Commander of Submarines North Sea, because I did not know their cipher key.

After I had found out all that I drove up the Jade the next morning showing a long homeward bound pennant completely reassured about this matter, through the High Seas Fleet lying at anchor, who had taken parade formation at my appearance on deck, to the lock at Wilhelmshaven. Here, too, there was a big reception for me with music, etc. – My berth in the shipyard was decorated with flags and garlands. I now had to present myself back at Fleet and report on my trip and wisely took all my orders from the Naval Staff in Berlin.

The Fleet Commander, Admiral Scheer, asked me immediately how I came to sink the Dutchmen if I did not receive the radio message. I then described the entire process to him and showed him the Naval Staff's orders, according to which neutrals should be dealt with without leniency and also mentioned I had a special cipher key from the Naval Staff.

Admiral Scheer asked his Chief of Staff, Admiral von Trotha, whether the High Seas Fleet had also been given these orders. The Admiral said no and added the Naval Staff had expressly stipulated to command U21 on its way back from the Mediterranean to its arrival in Wilhelmshaven only from Berlin. Even if the Naval Staff informed all the headquarters of the Foreign Office's compromise and demanded in a very strict form that nothing should happen to the Dutch steamers it possibly overlooked notifying the only submarine that was under its command.

The High Seas Fleet reported the matter to Berlin and relieved me of the further prosecution regarding this incident. I myself have heard nothing more about it.[64]

[64] The Germany allowed several dozen Dutch ships to sail in the war zone unmolested in addition to other concessions.

Bandoeng

Eemland

Jacatra

Zaandjik

Noordwijk

Gaasterland

XVI.
The Kaiser's Monitors

At Headquarters

In the fall of 1917 I was on leave in my hometown Kreuznach. Just when the GHQ was there. Since I was of the opinion: "Never go to your house if you are not called"[65] and since I also knew it should not be ideal for the Emperor to speak to submarine commanders at the moment, I preferred to keep my presence there as secret as possible and to cultivate well-deserved rest.

But one day I could not avoid meeting some of the headquarters staff guard who expressed their astonishment that I had not yet been to breakfast with His Majesty. They did not want to know anything about my negative attitude and I was called by telephone the next morning at 11:00 am with the information that I was ordered to have breakfast with His Majesty at 1:00 pm.

Although I didn't like it at all, I had no choice but to throw myself into full war paint and start the difficult walk to the resort.

When I got there the guard came out and I was led into the heavily carpeted hallways of the resort, where I was asked the reason for my request in a quiet voice by a bowing footman in a long court uniform with a smooth-shaven, impenetrable face.

Since I was aware that no one was admitted to the Emperor without first having negotiated with the Chief of the Cabinet, I reported first to the Chief of the Naval Cabinet, His Excellency Admiral von Müller.

---

[65] Jocular expression meaning to avoid closeness with one's superior.

The footman led me silently to the elevator and when I got to the second floor there was also a footman who asked me very quietly about my request.

Finally with his Excellency von Müller.

"Would you please tell me what you have done in the war?"

I was a little upset. Because as a Knight of the Pour le Mérite one could reasonably assume that a naval officer and adviser to the Emperor was informed of such matters.

I listed the engagements, which I didn't like very much, because it smelled of self-praise.

"Fine ... and what did you get the Pour le Mérite for?"

When I told him that I had received the medal in May 1915 for my trip to the Dardanelles, Müller interrupted me and said that he was now in the picture. He only asked me whether I had sunk warships there and who they would have been.

Then came a strange request: "Please tell me first everything you intend to tell His Majesty ..."

Then I was relieved. Now I went to the aide-de-camp adjutant on duty, who had only recently returned from the front and was happy to be able to speak to a fellow comrade again. We passed the time until the start of breakfast with a cigarette over the frontline stories.

When the time came we went into the anteroom where the court nobles met, all in field-gray uniforms. Also, the Chief of the Naval Cabinet was in field-gray!

I was the only one in a blue naval uniform.

When His Majesty entered all the nobles positioned themselves around in a semicircle and made a deep court bow. When the gentlemen held themselves up again I was introduced to His Majesty by the Chief of the Naval Cabinet, who was in the picture about my Dardanelles trip. We then went to the breakfast table where I had to take a seat to the left of the Kaiser. To my left was the Cabinet Chief, who watched every word I said to the Emperor.

When I told something about the front that the Emperor apparently should not know, I got a jab in a rib joint from His Excellency von Müller. I heard him whisper: "His Majesty doesn't care!"

Breakfast was very simple. There was a bouillon soup with meatballs and a cheese platter afterwards. With wine and champagne for the Kaiser.

When a second glass was poured the Emperor, I heard His Excellency's footman whisper in Müller's ear: "His Majesty is drinking a second glass of champagne!"

Then the Cabinet Chief to the footman: "Quickly report to His Majesty that coffee is served in the veranda."

Immediately afterwards the Emperor duly stood up from the table, regardless of whether one had finished eating or not and went to the porch, where he still had conversations with some gentlemen, especially with the Chief of the Naval Cabinet.

I was no longer honored by another speech, only when I said goodbye did His Majesty shake my hand and said he would soon like to go to the front again.

Since I had been with the Emperor, I did not want to miss visiting the working headquarters and went to Oranienhof, where I reported to Field Marshal von Hindenburg and General Ludendorff.

General Ludendorff received me immediately. Field Marshal von Hindenburg apologized for being so busy at the moment but invited me to his villa for dinner.

I was there promptly at 8 p.m. and met the gentlemen from the Operations Department who dined with the Field Marshal every evening. After everyone was assembled the double doors opened again and the Field Marshal, whom I saw for the first instance on this occasion, entered in his great, imposing shape.

He immediately came up to me and greeted me with a few words of apology that he could not meet me in the afternoon. He took my arm and went over to the dining room with me. I had to take a seat opposite the Field Marshal; General Ludendorff sat to my left and Colonel Bauer[66] to my right.

After the soup, the Field Marshal offered me a plate of peaches, I had to draw into a large glass. Afterward the orderly poured champagne into it, the Field Marshal held a short toast to me and alluding to my Dardanelles voyage at the end: "Gentlemen, I drink to the health of the only naval officer who brought about a decisive change in the World War! Cheers, my dear brother. "

The food was very good and simple. A strong vegetable soup, roast venison and fruit for dessert. After dinner we sat comfortably for a brief hour with a glass of wine on the veranda, the Field Marshal listening with interest to my stories about submarine voyages.

This was my first and last visit to the headquarters.

---

[66] Max Hermann Bauer, a protégée of Ludendorff.

XVII.
Fatal Errors

The submarine war would have defeated England!

The purpose of submarine trade warfare or unrestricted submarine warfare was to force England to peace by destroying its merchant fleet. England is an island and unable to feed itself. England required certain amount of ship tonnage or freight capacity in order to transport the necessary foodstuffs to the island. If England no longer has enough ships to bring sufficient food and raw materials to its island it must either starve or make peace.

The submarine war also greatly reduced British fighting power on land. England had to withdraw a great many men from the Front to keep her yards in full operation, to replace sunken steamers with new ones, to build all her innumerable new vessels for anti-submarine defence, she needed men to man these vessels, she had to run her coal mines and munitions factories, yes, she had to still further increase the capacity of these establishments since she lacked the necessary supply from America due to lack of shipping space, etc. When the intensified submarine war was about to be declared, we submarine commanders were asked how many thousands of registered tons could we guarantee each month based on the current level of enemy countermeasures.

We replied that we would probably sink 600,000 tons a month, provided submarine losses were immediately replaced by fully commissioned new construction.

I just want to mention here that at the beginning of the war we had only 20 submarines ready for the front. Several hundred submarines were built during the war; we also lost 199 submarines during the war. Had the intensified submarine war started shortly after the beginning of the war, the war would have come to an end much sooner, despite the lower number of submarines, because at that time the enemy was not yet up to the task with well developed countermeasures and accordingly our losses would have been significantly lower. But we waited until the enemy had fully expanded his countermeasures because, to use the words of our Chancellor Bethmann-Hollweg, we "only wanted to wage war humanely". The more intense war is, the more humane it is because it ends faster. In this case, millions would not have bled to death on the front lines and many thousands would not have died of hunger in the homeland.

Thus in February 1917 the intensified submarine war began and submarines dutifully sank the promised 600,000 tons per month. Even more was sunk every month. In two months it was over a million. The intensified submarine war had its effect. The enemy was ready to fold and ask for peace a year after the intensified submarine war began.

The Englishman took the last rigorous measure that was left to him by seizing all the neutral merchant shipping that he could get hold of, be it in his ports or on the high seas.[67] Neutrals he bought off with a sum of money. But as so often during the war, when we had almost forced the enemy down at the Front, there was always an enemy ally behind our own front at home. Be it the Reichstag with its peace resolutions or munitions workers' strikes staged by foreigners. It is like a curse that weighs on the German people from the old Nibelungenlied saga repeating itself over and over and Siegfried is always treacherously slain by the black Hagen.

Despite all of this, the intensified submarine war continued and submarines continued to sink.

Now one or the other of my readers may say here, how was it possible the American came over with his troops and his war

[67] Most likely referring Great Britain's contracting Norwegian shipping in early 1917 shortly after the German declaration of unrestricted warfare.

equipment when submarines were sinking so much? I can only answer that it was impossible to shut off the whole ocean and how hard the enemy made it for our submarines to sink their merchant fleet is clear enough from the descriptions of their countermeasures. In addition, it was not possible for a submarine who could only see the steamer through the periscope before it was torpedoed, determine what the steamer had loaded, whether ammunition or troops, war material or food, which ultimately did not matter to us submariners. The main thing was we sunk the enemy's cargo capacity in order to force him to his knees in this manner

And so it was. The enemy could not keep up with our sinking numbers in his new construction and we would have led the war to a victorious end if we had not been prevented from continuing. According to genuine news which came to us from England after the armistice and is also confirmed by the memoirs of enemy statesmen, the intensified submarine war would have led to a victorious end for us if it had lasted 1-1/2-at most 2 months more.[68] There Hagen, there the revolution brought to Germany the shameful armistice conditions and betrayed our victorious and courageous Submarine Force.

[68] A sentiment demonstrated in a number of German submarine commander memoirs.

XVIII.
Why the fleet was not used

Misses opportunities...

It was once said the only reason we did not win a final victory was because the war was nothing more than a series of missed opportunities. If, after the end of the war, one looks soberly at all the events, one could almost assume at least as far as naval warfare is concerned, this statement is quite correct.

Why didn't the entire High Seas Fleet deploy against England at the beginning of the war, as Admiral von Tirpitz always insisted upon? It was later said the non-deployment of our fleet was due to a letter from the then Director-General of the Hamburg-America Line, Albert Ballin, to the Emperor, in which Ballin is said to have asked the Emperor not to use the fleet, otherwise England would be "even more irritated" and our trade and our overseas merchant ships would then be confiscated.[69]

The truth is that at the beginning of the war the English fleet was not yet equipped to successfully face our German High Seas Fleet Only in 1916 did the Englishman think himself so far superior to be able to destroy our fleet. But despite all this, the Englishman in the naval battle off the Skagerrak, despite almost doubly superior numbers, was unable to best our High Seas Fleet. On the contrary, he retreated during the night so as not to suffer any more losses.

[69] In the years leading to the Great War and up to July 1914, Albert Ballin attempted to broker a naval and maritime agreement between England and German.

In the course of the war, the Englishman no longer dared to face our High Seas Fleet in battle on the open seas.[70] It is understandably human that under these circumstances the morale of the crews of our ships was worn down during the constant exercises and lying around in the roadstead, and under these circumstances the poison of revolution could find their way more easily with them than with frontline vessels which were constantly at sea and came into contact with the enemy.

Another missed opportunity is the too late deployment of the intensified submarine war which is due to too much consideration of our diplomatic leadership towards the enemy and neutrals and above all the fear of false reports in the enemy press regarding our barbaric warfare. When there is war, it is always a matter of life and death, and if I have a weapon with which I can force the enemy to the ground, I also use it and do not wait until the enemy has built up armor that protects him from my weapon. The more intense war is, the more humane it is, because the faster it ends. If we had started the intensified submarine war immediately at the beginning of the war and deployed our High Seas Fleet towards the enemy, we should have had peace by 1917 at the latest and hundreds of thousands would not have had to bleed to death at the front and at home starving to death from the hunger blockade against Germany carried out in violation of international law.

[70] A surprising assessment by Hersing considering he indicate access to postwar sources which clearly state otherwise.

XIX

Could the revolution have been crushed?

Panicked leadership - My submarine flotilla firmly in my hand

I do not want to judge whether the revolution could have been quashed if the authorities at home intervened in good time, I just want to talk about my experiences during and after the revolution.

At the beginning of November 1918, my good boat U21 was at the Imperial Shipyard in Kiel for major repairs and overhauls that were to take several weeks. During this time, I was the half-flotilla commander at the submarine school and, with my many years of experience in the front, was supposed to train the newly commissioned submarine crews there before they went to the Front. On Sunday, November 3, I was in the port of Kiel with my support ship, the steamer "Lucie Wörmann".

Suddenly a non-commissioned officer: "Lieutenant, the alarm has sounded in Kiel ..."

"What's going on?"

No one knew anything specifically.

In the evening personnel arrived from Kiel: "Now the alarm has been canceled." Otherwise they don't know much either.

I inquire at the Torpedo Inspection Bureau by telephone. Received all kinds of information. Confused, uncertain fumbling, unclear ... it was nothing to do with them "nothing else special happened ... the unrest of the times ..."

Monday morning I let all the submarines run out. I intended to go out later with the steamer.

Suddenly a message, a radio command ordering the immediate return of all submarines ...

Oh ... now I knew more than yesterday. I immediately gave the order: all incoming submarines to go alongside the support ship. All contact with shore is strictly prohibited ...

So I held my strength tightly in my hand ... and a dozen submarines were a force to be reckoned with.

I race ashore, race to the [Torpedo] Inspection. Must know what the hell was going on ...

Dumbfounded faces at the Inspection.

I was certainly not polite: "For Heaven's sake someone must have given the order ...? "

In response, a poor one: "We don't know ..

To the Admiral.

He also does not know who gave the order to return.

That was a fine state of affairs ...

I receive a message that all my submarines have come in safely and have gone alongside the support ship. The Inspection is asked for further orders.

I get one: it is the strangest command and at the same time the most devastating I have received in my entire career. It reads: "Do what you want!"

"Thank you," I bark, "that's an order too ..."

I am off to the boats.

I immediately command to depart and follow with the support ship.

At this departure we pass the 4th Squadron with its heavy ships lying powerfully at their buoys...

I felt an uncomfortable tingle. What about the crews in the 4th Squadron?

If they prevented us from leaving?

Then there would be a battle ...

But something completely different happened. Something quite frightening. The big ships slid past us, slid astern of at us ... nothing stirred in their superstructures ... and no one on deck ...

Deserted? I couldn't explain it ...

The next dreadnought ... deserted ...

I later found out. The crew had fled for fear of us. We were considered "unreliable".

Rightly so!

In the evening at around 8 p.m. I ran into Eckernförde with my boats and moored there at the submarine pier.

While the ship was still running in the cruiser "Dresden" passed me at high speed and also anchored in Eckernförde. I immediately tried to obtain more information by telephone from Kiel to plan my future actions. At first I only received confused reports until the telephone was completely interrupted around half past ten in the evening.

The commander of the cruiser "Dresden", HRH Prince Adalbert of Prussia, asked me not to send any crew on shore leave because there would be revolution in Kiel. However, I had already let my personnel go ashore and everything was quiet in Eckernförde; I saw no reason to order my crews back on board. Above all, I didn't want to spread unnecessary unrest. I also knew very well that the submariners would never give into revolution. They all returned from shore leave on time.

Around midnight several engineering trainees suddenly came on board to me from Kiel to Eckernförde and I now found out what was going on in Kiel. The engineering trainees had fired intensely against rebels in front of the deck officer school in Kiel-Wik and afterwards, because they could not resist the superior force, were ordered to disperse by their superiors instructing them to come to me in Eckernförde, otherwise they would be lynched by the rebellious crowd. In Kiel itself, in the garrison as well as on the ships, soldiers' councils had formed and seized power. There was nothing more to be done there for the time being. I immediately called my commanders to a conference to keep them informed of the events.

In the middle of the night there was news: the rebels were zooming up in trucks to get their hands on the submarines. They have many machine guns with them.

I thought quickly. If I dared to fight: how would the "Dresden" crew behave?

I knew that I could finish the rabble and their machine guns in two minutes. Naive, an attack on my boats. But the "Dresden" with its heavy guns?

I had no answer to one question: was it a local phenomenon or was it fermenting throughout the empire?

I called the Prince: "How would the Dresden's crew behave?"

The answer was not encouraging: "I can actually no longer predict with some certainty ..."

Then I ran out with all the boats to hold – Lord God of Mannheim![71] as it was – fighting power firmly in my hand.

When departing last in the support ship from Eckernförde Bay we just saw the first trucks racing in at dawn. Red scraps fluttered in the wind ...

We saw through glasses how they took machine guns off the trucks. Heard the crackling of M.G.s. But there were no impacts from his tiny water columns in the bay ... it was childish to try to shoot at me at four kilometers with those squirts ...

I drove my boats east constant exercising them and initially tried to get in touch with land in the Bay of Lübeck. But revolution had already broken out there. Now I went on to Swinemünde, which according to reliable news, should still be completely quiet in order to take on provisions again since the boats only had provisions on board for a few days during their training phase.

---

[71] Expression of amazement common to parts of Germany.

When I entered Swinemünde from a distance I saw two cruisers lying at the pier; but as intensely as we looked with the best binoculars glued to our eyes out we could see neither flag nor pennant. When we drove past them - it was the cruisers "Regensburg" and "Dresden" - we only discovered a small watch on board of home guards from the artillery regiment in Swinemünde. The cruisers' crews had left their ships taking the ships' fund, ran to the train station, fired up locomotives, assembled together trains and gone to Berlin. Here in Swinemünde there was still complete calm and the commander's office was still in perfect order, I could now calmly wait for the course of events and collect news. From here I also had direct telephone connection to Berlin and was constantly provided with solid news. Now came the hardest days for us that an old combatant and loyal soldier who loves the fatherland can experience.

On November 9th the Emperor's abdication and then the worst, the conclusion of the shameful armistice.

From Berlin I was ordered to bring the boats to Kiel and make them ready for delivery. At the same time, I was only given provisions for the duration of the crossing. Since I realized it was useless to resist this order with my submarines, especially since the armistice had been signed, and I had also learned other very sad news from the Reich regarding the retreat of the Western Front, with a bleeding heart I finally had to submit and start the crossing to Kiel.

I first contacted Kiel by telephone and conferred with the then commanding officer in Kiel, Mr. Noske.[72]

I informed him I had orders to come to Kiel and wanted to ensure neither the boats nor the crews would be fired at, otherwise I would reply with a very heavy caliber.

[72] Gustav Noske, a Reichstag deputy of the Social Democratic Party. He was sent from Berlin to assume the Governor of Kiel office. The Soldier's Council welcome his arrival.

Noske solemnly promised to keep things orderly. But when I explained to him that my boats only came in under the old battle ensign and there would be no red flag seen from me, he did not want it and absolutely demanded the red flag.

I remained firm however and he finally gave in so as not to endanger the delivery of the submarines. The immediate delivery of the submarines was one of the main conditions of the shameful armistice.

I now drove my submarines back to Kiel where the boats were moored in the shipyard and officially decommissioned. Before that the individual crews were photographed again on their boat beneath the old battle ensign. The crews themselves were properly rewarded and sent on leave for the time being. Let those who were in favor of the shameful armistice see for themselves how the boats were to come to England.

I myself moored to the buoy with the support ship to carry out their processing. If the port of Kiel had made a bleak impression the last time I left, it was now no longer recognizable. The formerly proud German fleet was partly in the port of Kiel, a red flag at every tip of the mast and yardarm, alongside large barges into which their ammunition was unloaded. Not a single boat crossed the harbor that did not have at least one red flag! One had to shake in disgust at such a picture! I want to tell you about a little experience that really sheds light on the situation at that time.

Of course I had at the top our old battle ensign waving from my support ship and the half-flotilla standard as my command flag. The various soldiers' councils could not say anything against the battle ensign because Noske allowed this flag to be flown, but seemed to be more annoyed with my half-flotilla standard.

On a fine noon day a signalman came to me and smiled and presented me with the semaphore signal log in which the following signal from the large battleship next to me was written: “Soldier's Council to Soldier's Council. Please promptly bring down the command standard. "

Since there was no soldier's council on board the signal was directed to the ship, I had to answer the signal and as polite man had the following signaled back: "Half-Flotilla Commander to soldier's council. Somebody shot you in the brain ... and forgot to stir. "

I received the following answer again: "Soldier's Council to Half-Flotilla Commander. The last sign was official. "

As I was recognized by the soldier's council as the half-flotilla commander I now signaled back as an answer: "Half-Flotilla Commander to soldier's council. My answer was also official. I will not deal with you otherwise. "

According to this information, I was no longer bothered by anyone and my flag and standard remained in peace.

Submarines and cruiser in background at Kiel in November 1918. Red flag flying at middle right.

XX.
The Bolshevik plague in Riga

With the Iron Division – A Burning German Theater - Brutish people loot, burn, disgrace, murder

After I finished delivering my submarines and relieving the crews at the beginning of December 1918, I also returned my support ship, the steamer "Lucie Wörmann", to the Wörmann Line shipping company in Hamburg from which the steamer was chartered. I returned to the station commander in Kiel and asked for further orders.

At that time naval officers were scarce because most of them had been on leave at the beginning of the revolution. Finally, I was ordered to embark for Riga to oversee the orderly loading of troops at the Russian front. As much as possible of the provisions stored there should also be transported to Germany. I took the steamer "Hannover", which was the first to depart to Libau and from there through the Russian mined area to the Gulf of Riga, to Dünamünde, then further up the Daugava to Riga.

In Riga itself the Army Command, which lived in the old German knight's palace in Riga, was still completely in control. Only the troops arriving for shipment had absolutely no discipline left. They had been at the front line at every phase since 1917 and had become completely Bolshevik. When the temperature dropped to 18 degrees they came in individual groups, mostly without officers, pounded by the snow, of course drunk, and stormed the next steamer without permission in order to get home as quickly as possible. They left horses and wagons stuck somewhere in the snow and refused to feed the animals. Entire field guns with horses were sold for 200 rubles and several demijohns of rum. Napoleon's withdrawal from

Moscow more than 100 years ago could not have been worse. With a lot of effort and hardship and with all my energy, I succeeded in sounding out the soldiers' councils and reinstating some officers back into their positions so a controlled evacuation was possible. I was supported in every way by the Army High Command, but eventually the Army High Command was powerless against the mass of men.

Now that I had brought some discipline into the troops by threatening them to freeze and starve to death in Riga, I was finally able to send around 18,000 men and eleven steamers to Germany by the beginning of January. But it was high time too, because the Russian Bolsheviks were pressing ahead with all their might.

The Reds were soon driven out again by the "Iron Division", which was forming in the Baltic States and could not be shipped under any circumstances. Instead, it was to be accommodated on the estates of the Baltic barons after a victorious campaign in the Baltics.

On December 31, 1918 I had just returned from dispatching a steamer from the departure port and wanted to receive new information and orders at the castle.

Strange: there was no sentry post in front of the castle. I went into the castle, it was deserted. Finally I met a sergeant with an "Iron Division" insignia. He informed me the Army Command had ordered automobiles this morning and departed to Germany. But there was still a letter for me. He rummaged in his pockets. Found the command.

It was very laconic: "The High Command is returning to Germany. Send as many troops as possible and try to get to Germany with the last steamer. "

I finally found an officer of the "Iron Division ". He described the situation as terrible. He did not believe the troops would be able to stay in Riga beyond January 1st or 2nd. They would fall back over the Daugava River.

The most fantastic New Year's Eve I've seen: the one at the turn of 18 to 19 ...

A pillar of flame rose from the city of Riga, people in the harbor were racing around ...

"Riots in the city ...", people called to me and God knows who else ...

"The German theater is on fire ..."

Then I heard a strange song that I hadn't heard in a long time. Big parabolas whizzed over the city ... shells. Russians firing on Riga.

The rabble in frantic activity. The rear of the "Iron Division" who opposed the Russians with their machine guns hailing towards them, the homes of the Germans were looted, parades with red flags pushed through the streets, some students were making speeches, but at the rear, the large crowd became ever thinner, the horde scattered. Defenseless German women were attacked, robbed and murdered in their apartments. It was New Year's Eve today, it was a great spectacle ...

Slowly, step by step, the "Iron Division" withdrew ... chaos followed in the vacuum: murder and filth, fire and turmoil ...

I went down the Duna estuary with my icebreaker "Hindenburg" ... I wanted to take up two battalions in Dünamünde ...

The port looked terrible. The soldiers, gloomy, weary, looking despondent under their steel helmets, women in light clothes, a few things in their hands ... old men, trembling from the effort of walking ...

In German they begged me to save them, to take them to Germany ...

A beautiful woman, her hair loose, jewelry clenched around her hands, shaking by crying ...

A boy talking crazily ... mother murdered, father dragged away...

It was a long time before they all realized that it was serious when I said they could go to Germany ...

The roofs of Riga glowed red in the distance ... a terrible evening red of fire and smoke ...

I stand on the bridge ... the gangway is hauled up ... the ropes are released ...

Screaming on the quay ... fills the evening. Strangely animated people ... ladies in thin silk bunting, ridiculous webbed silk stockings . . . Men whose pathos sounded eerie ...

"Take us with ... it is all over ... our theater is on fire ..."

So I took on board the ensemble of the German theater in Riga, while the last charred beams collapsed in the theater and the mob on the street draped themselves with paper crowns and tinny Hamlet swords ...[73]

As the "Hindenburg" slowly left the port, Riga filled with the Bolshevik troops and the new "government" began to "work", and the first "positive measures" were taken. Limestone was crumbled from a wall by jurors who murdered Germans ...[74]

---

[73] The German theater was considered the highpoint of German culture.

[74] Germans who stayed behind referred to the Bolshevik occupation as the "Time of Horror". Thousands were reputedly executed. Riga was retaken by the German organized Free Corps in May 1919.

XXI.
The Ghost ship

U 21 goes to last rest

Only when I was back in Germany did I fully understood the bleakness of our situation at the time. The effects of the ceasefire agreement were terrible. Our naval bases empty. The once proud, powerful and undefeated German fleet had gone to England and was interned in the port of Scapa Flow. Most of our brave submarines have been delivered to England; only a few pitiful remnants that could no longer sail lay at the shipyards.

Including my brave U 21.

But the enemy was adamant; every submarine that could be patched up had to be delivered.

So U 21 was now made ready to be handover, but the engines no longer wanted to get going. Finally the boat was to be towed. At the beginning of February 1919 the transfer to England took place without crew in tow of the tugboat "Laboe".

The trip first went through the Kaiser Wilhelm Canal and then into the North Sea, in very bad, rainy and hazy weather.

The ailing captive boat followed its escort ...

The nighttime voyage of February 21st to 22nd was scary.

The boat travelled down the same route it had once crossed as a winner, a fighter, on brighter nights, in a more beautiful world ...

The boat, the abandoned, deserted boat, dreamed of those bright nights, dreamed of the German sailors, dreamed of the officers in the tower ... dreamed of how quickly it dived ... how it stole past the patrols ...

No one in the tower today, nobody. Nobody at the engines, nobody at the control room ...

No one at the stations: a ghost ship ...

But it was not unmanned, the boat, its soul was not dead. The soul touched levers and wheels, the soul was at diving stations ...

It was an eerie ghostly night and the sea was wild. The sailors on the tug were not comfortable on this trip ... The seas threw itself against the ship, white spray jumped up at the bridge ...

Now bad weather ... the was haunting ...

Suddenly the sailors altogether were hurled on the deck such that they had to find their aching limbs. A terrible jolt shook the ship as if the devil was holding it aft ...

Then the crew stared: the steel rope sang with pain, no longer hold able to hold its weight ...

"The submarine is sinking ..." they roar.

The captain swears ... what should he do, the bitch of a submarine pulls him to the bottom ...

"Let go the rope ..." he shouts ...

Silently, detached from his tormentor, detached from this world, in silent pensiveness U21 glides down into the silence from which there is no more awakening ...

As if with healing, caring hands, the sea depths welcomes its loyal comrade ...

At Latitude 54 degrees 19 minutes North!

Longitude 3 degrees 42 minutes West!

The most successful boat, U 21, lies on the seabed there. Rest gently in peace in the infinity of the sea ...

Made in the USA
Monee, IL
04 November 2026

67384900R00075